Project Management

How to be a Successful Project Controller?

By
Kadidia Fofana

Table of Content

CHAPTER ONE

INTRODUCTION TO PROJECT MANAGEMENT

BACKGROUND
WHAT IS A PROJECT?
  1. *Temporary nature*
  2. *Unique products and services*
  3. *Progressive elaboration*
WHAT IS PROJECT MANAGEMENT?
WHAT IS THE PROJECT LIFE CYCLE?
PROJECT PROCESSES
PROJECT LIFE CYCLE PHASES
  Initiation Phase
  Planning Phase
  Execution Phase
  Managing and Controlling Phase
  Closing Phase

CHAPTER TWO

PROGRAMS AND PROGRAM MANAGEMENT

PROGRAM LIFE CYCLE PHASES

CHAPTER THREE

PORTFOLIOS AND PORTFOLIO MANAGEMENT

PORTFOLIO PROCESSES
  Standards
PORTFOLIO LIFE CYCLE PHASES
  Inventory Phase
  Analysis Phase
  Alignment Phase
  Management Phase
  Closing Phase

CHAPTER FOUR
PROJECT STAKEHOLDERS

    Types of Stakeholders
    Key stakeholders within the project include:
      External customers
    Organizational Influences

CHAPTER FIVE
THE PROJECT CONTROLLER

    Emergence of the Project Controller
    Key General Project Controller Skills
      1. *Leadership*
      2. *Communication*
      3. *Negotiation*
      4. *Problem solving*
      5. *Influencing the organization*
    Project Controller Responsibilities
    Types of Project Controller
      1. *The Project PCO*
      2. *The Program PCO*
      3. *The Portfolio PCO*
    Why Organizations Need Project Controllers
    Role of a Project Controller in Budget Management
      1. *Properly understand the stakeholder's needs and wants*
      2. *Budget for surprises*
      3. *Develop relevant Key Performance Indicators (KPIs)*
      4. *The "Three Rs"*
      5. *Keep everyone informed and accountable*
    Defining Proper Project Control
    Principles of Project Control
      1. *Clarify project baseline*
      2. *Control project cost*
      4. *Proactively evaluate progress*
      Five Effective Project Control Practices

1. *Holding meetings*
2. *Performing quality control*
3. *Tracking progress*
4. *Responding to changes*
5. *Managing risks and issues*

CHAPTER SIX

PROJECT MANAGEMENT METHODOLOGIES

- *Waterfall*
- *Agile*
- *Hybrid*

CHAPTER SEVEN

COMMON PROJECT MANAGEMENT MISTAKES

Failure to prioritize tasks and projects
Forgetting that project management is also people management
Failure to communicate regularly with the team members
Allowing changes get out of hand
Failing to use a project management tool
Not adjusting the course of the project when things go wrong

CHAPTER EIGHT

TAKE CONTROL OF YOUR CAREER

TRAITS OF AN EXCEPTIONAL PROJECT CONTROLLER

REFERENCES

CHAPTER ONE

Introduction to Project Management

Background

Project management is one of the fastest growing professions. Like other professions, including medicine, law, and accounting, project management has an accumulated body of knowledge that is built and advanced by its practitioners and academics. You might be wondering, "What does a 'body of knowledge' include?" A set of proven and widely applied traditional practices, as well as information and guidelines for innovation and advancement, make up a body of knowledge. Project management's body of knowledge is presented in detail in The Project Management Body of Knowledge (PMBOK®), a guide offered by the Project Management Institute (PMI).

According to the PMI's "Job Growth and Talent Gap" article, the disparity between employers' workforce needs and the availability of project management professionals is continually widening across the globe. This trend was introduced by the PMI's first talent gap analysis in 2008, and has since outpaced 2012 analysis projections. There are a number of factors that may close the gap. These include a dramatic increase in the number of jobs that require project-oriented skills, attrition rates (such as an increased number of professionals retiring, which opens up job opportunities), and an increased number of people with a project management skill set. For example, in China and India, there is a significant uptick in the number of qualified project professionals. These factors reinforce the ability of project professionals to introduce change and innovations to numerous organizations. Indeed, projects have the ability to change the world, both directly and indirectly. As a result, project professionals of today and tomorrow are at the forefront of amazing opportunities, hence the imperative to motivate more people to join this profession.

Did you know that the rate of project failures is high? Yes according to the PMI in the annual survey 2018 "More than half of IT projects are still failing".

Based on this, you might guess that organizations are content to settle for only a moderate level of project success. However, this is not the case! Despite the challenges that project professionals face, organizations expect their projects to be completed faster, better, and cheaper.

In this book, we will delve further into the topic of project management. To begin, we will keep the concept simple: in order for project targets and goals to be reached, it is important to ensure that available resources follow a set schedule of use. While both project managers and project controllers are involved in keeping project team members on an established course, the project controller has a set of responsibilities distinct from those of the manager. We will further explore the difference between a project manager and a project controller in subsequent chapters, and Chapter Five explores the project controller position in depth. To avoid confusion, when talking about concepts relating to both managers and controllers, we will use the term "project professionals," or "project leaders." As many books provide guidance for project managers, but not project controllers, this book aims to be a reference primarily for project controllers. Other project professionals seeking success in their careers may benefit from reading this book, as well.

What Is A Project?

An organization performs a wide range of activities defined by its legislature. These activities take the form of either projects or operations. Projects and operations may overlap in some cases, and often share a number of characteristics. Both are performed by people, both are constrained by limited resources, and both are planned, executed, and controlled.

An **operation** simply refers to ongoing, repetitive activities. Operations lack defined beginnings and endings, and the products and services produced may not necessarily be unique. The objectives of operations are aimed at sustaining the business.

On the other hand, a **project** is a temporary undertaking aimed at creating unique products and services. The work is "temporary" in the sense that there is a definite beginning and ending, and the products and services are "unique" in the sense that they have features distinguishing them from all other products and services. Projects often fulfill needs that organizations cannot address by means of their normal operations.

Projects often involve multiple levels of an organization. They can involve one person, or thousands of people. Their temporary nature does not mean that they are short-term; they can be conducted within a few weeks, or they may span 5 years. Additionally, projects may involve one organization, or may link a number of organizations through partnerships or joint ventures. Projects are a means by which an organization implements its business strategy. Examples of project goals include:

- ✓ Developing a new product or service
- ✓ Effecting a change in structure or organizational style
- ✓ Launching a new enterprise

The three defining characteristics of projects are its temporary nature, unique products and services, and progressive elaboration.

1. Temporary nature

As mentioned earlier, a project is "temporary" because it has a definite beginning and a definite ending. However, you might ask, what defines the end? Well, a number of different circumstances can result in the end of a project. In one scenario, the project comes to an end when the set objectives have been achieved. In a second, the project ends when the set objectives cannot be met or realized. In a third, the project terminates when the need for that project no longer exists. The important thing to understand is that projects are never ongoing, open-ended efforts.

In addition to referring to the project's limited duration, "temporary nature" also refers to the market window of opportunity, and to the project team. The need for the project may arise only during specific and transient market conditions. Most organizations often set up a team solely to conduct a project, and once the project is closed, the team is disbanded along with it.

It is important to note that while the project itself is temporary, the products and services created by the project may have ongoing effects. The project may lead to both intended and unintended social, economic, and environmental effects felt long after the project concludes. For instance, a three-year project to develop a vaccine against Ebola may create results that last for centuries.

2. Unique products and services

A project does something that has not been done before; it is unique. What if the category to which a project belongs is large? Will the product still be unique? The answer is yes. For instance, you might look at a building under construction and wonder, "How is this unique, when there's a building just like this right across town?" Well, this building is unique because it belongs to different owners. It has a different architectural design, and it is located on different land. It has a different construction company, and may display other distinguishing traits. Despite having comparable features to other similar products, this product is still unique.

3. Progressive elaboration

Progressive elaboration is a project characteristic that we have not yet discussed. Progressive elaboration is the continuous improvement and embellishment of distinguishing features. "Progressive" means that the process occurs in steps or increments. "Elaboration" means that distinguishing details receive extreme care and attention. These distinguishing details are identified early in the project, and become more explicitly defined as the team continues to develop a better understanding of the desired product. The team must carefully coordinate progressive elaboration efforts so that the work remains within the limits of the project scope, especially if the project is conducted under a contract.

What is Project Management?

Now that we understand projects, we can define project management. **Project management** is the application of knowledge, skills, tools, and techniques to project activities, in order to realize a project goal. It is the role of both the project manager and project controller to balance competing demands (time, cost, scope, risk, and quality), identify requirements, and deal with stakeholders, who may have different expectations and needs. (We will define "stakeholders" more precisely in Chapter Four; for now, consider them clients, sponsors, and project professionals.) These management tasks are often iterative; in other words, they are often carried out repeatedly throughout the life cycle of the project.

Project professionals must understand that projects are bound by three integrated parameters: time, scope, and cost. This is known as the **triple constraint principle**. Professionals often use the triple constraint principle as a framework for evaluating competing demands. This principle is often depicted as a triangle, with each side or point representing a parameter.

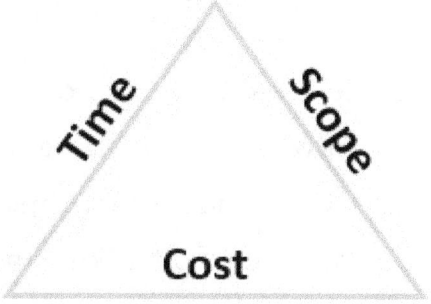

Figure 1. The triple constraint principle.

A change in one parameter requires tradeoffs in the others. For instance, when changing the scope of the project, there will probably be an impact on cost and time. Improving performance in one unit may require sacrificing performance in another unit. In some cases, the interaction between parameters is straightforward. In other cases, the interaction is subtle, yielding a number of uncertainties. For a project to be successful, the project professional must ensure that these tradeoffs are balanced and well managed.

What is the Project Life Cycle?

Since the undertakings of a project have some degree of complexity, most organizations divide a project into phases to make management easier. Collectively, these phases are called the **life cycle** of the project. The major phases of the life cycle are **initiation, planning, execution, controlling,** and **closing**. We will discuss these in detail momentarily.

Each phase is delineated by the completion of one or more **deliverables**. Deliverables are tangible and verifiable work products that help divide the project into a logical sequence. The deliverables offer a tangible way to continually assess the overall progress of the project. Typically, the deliverables of one phase determine whether the project continues to the next phase. They enable managers to detect errors mid-project and take corrective action. Deliverables are assessed during **phase-end reviews**, which take place at the end of each phase. They are often referred to as **kill points, stage gates,** or **phase exits**.

Phase-end reviews usually involve some form of technology transfer or handoffs between team members. Before the next phase begins, the deliverables from the previous phase must be approved. However, work on the next phase often starts before approval of the previous phase. This occurs when the benefits of continuing work prior to approval outweigh the risks. This practice of overlapping phases is referred to as **fast tracking**.

Most projects have similar life cycle phases and deliverables, but very few of them are identical. Many projects have four or five phases, but others may have nine or more. Additionally, sub-projects that lie within a larger project can have different life cycles. Be careful when distinguishing between *project* life cycle and *product* life cycle

Project managers must provide a description of the life cycle before the project begins. This description may be general or detailed. Detailed descriptions include a wide range of forms, checklists, and charts, offering a documented structure to the project. Detailed descriptions are known as **project management methodologies** (PMMs). PMMs define the technical work to be done in each phase, and the team involved in each phase. There are different types of PMMs, and these are discussed in more detail in Chapter Six.

There are two common life cycle trends that PMMs must take into account:

1. The probability of a project reaching successful completion is lowest at the start of the project, when the levels of uncertainty and probabilities of risk events are at their highest. There is a progressive increase in the probability of successful completion as the project advances. Therefore, it makes sense to begin with low cost and staffing levels, and increase these only as the probability of completion increases. Figure 2 illustrates this trend.

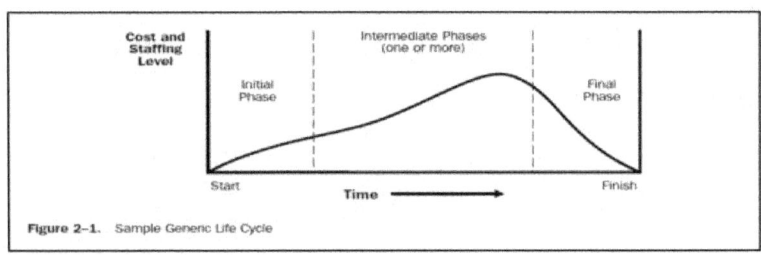

Figure 2. How cost and staffing levels progress from the initiation phase to the closing phase.

2. At the start of the project, stakeholders have a significant amount of influence over the final products and project budget, but their influence gets progressively lower as the project advances. This makes intuitive sense, right? Say your client decides he wants a third office in the building instead of a bathroom. If you are still designing the blueprint, this is a relatively simple adjustment. If you have already put the toilets in, this becomes a costly and time-consuming endeavor.

Project Processes

Each phase of the life cycle involves numerous **processes**. A process is a series of actions that yields a result. These processes are performed by people, and are categorized as either **project management processes** or **product-oriented processes**. Project management processes describe, organize, and complete the necessary activities of a project. Product-oriented processes specify and create the products of the project.

Project management processes involve **core processes** and **facilitating processes**. Core processes are basic processes that are repeated throughout all phases of every project. These include planning and defining the scope, defining the type and duration of activities, developing schedules, planning risk management, allocating resources, and estimating costs. Core processes often depend on one another, which requires that they be conducted in a certain order. For instance, you have to estimate the cost of an activity before you can allocate resources to it, and you have

17

to define the duration of the activity before you can schedule it.

Facilitating processes are processes that vary based on the individual nature of the project. They include quality planning, organizational planning, staff acquisition, communication planning, identification of risks, qualitative and quantitative risk assessment, and risk response planning. These processes are performed intermittently, on an as-needed basis. For instance, in some cases, there may be few or even no identifiable risks associated with the project at the beginning. However, soon after the planning phase is complete, risks associated with cost and time can appear and jeopardize the project.

Project Life Cycle Phases

We will now examine the major phases of the project life cycle, and the key processes of each phase. Remember, the phases are **initiation, planning, execution, controlling,** and **closing.**

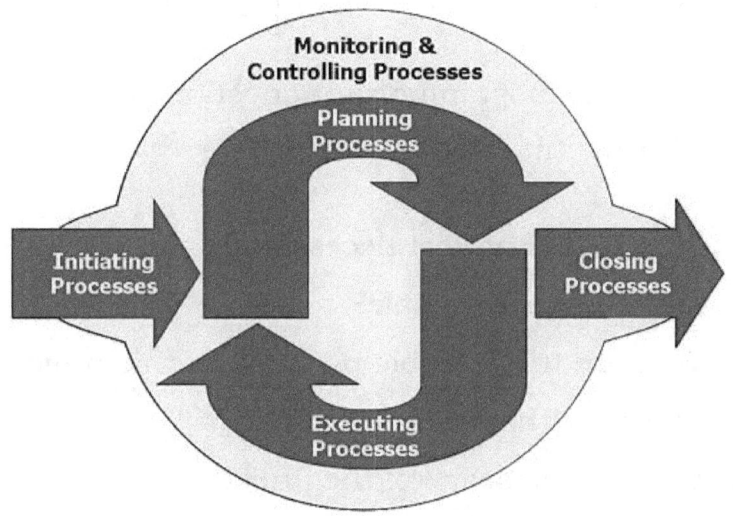

Figure 3. The project life cycle.

Initiation Phase

In this phase, the project is introduced and authorized. The idea of the project is evaluated. The most important goal of this phase is to determine the project's feasibility, and whether the project has a solid base of support. Decisions are then made regarding who will perform the work and which stakeholders will be involved. During initiation, prospective leaders draft a proposal describing the project. Two examples of proposal types are business plans and grant proposals. The sponsors of the project will then evaluate the proposal, and once all is approved, provide the necessary funding. Approval of the proposal officially marks the beginning of the project.

Some of the most important questions that need to be answered during initiation include:

- Why is this project necessary?
- Is the project feasible?
- Who are the possible partners in this project?
- What are the desired results?
- What is the scope of the project?

During the initiation phase, the partners in the project should form a temporary relationship. To avoid conflicts and false expectations, all parties involved must agree on the type of project. Project types include:

- research and development
- a project that will deliver a prototype or 'proof of concept'

- a project that will deliver a working product

Choosing the type of project largely determines what products and services are created. Consider, for example, a software company that wants to develop a new application. A research and development project might deliver a report assessing the technological feasibility of the app. A prototype project might present all the utilities of the app, with the understanding that they might not yet be functional for users. A project that delivers a final, working app must take into consideration matters of maintenance, instructions, and operational management.

It is important for the parties involved to clear up any areas of confusion during the initiation stage, in order to avoid future misunderstandings. For instance, the clients may expect a working product, while the team members think that they are just developing a prototype. A sponsor may be under the impression that the project can easily produce fully functional software, while the team members still need to examine the feasibility of the idea. Therefore, during initiation, all key stakeholders must do their due diligence in deciding whether the project is a go. Once the project is given the green light, they will now need to create a **project charter** or a **project initiation document**, which outlines the project's main purpose, constraints, requirements, deliverables, and evidence of approval.

Planning Phase

This is one of the most important phases of the life cycle. This phase involves a relatively large number of processes. The amount of planning that goes into a project is proportional to the scope of the project, and to the utility of the information that can be gained from planning. Unlike the initiation phase, which has a defined beginning and end, this phase is ongoing, and runs throughout the entire life cycle. Many planning processes must continually be revisited prior to project completion. For example, if the initial completion date set for the project needs to be changed, then plans for budget, scope, and resources may have to be reviewed as well. Concrete plans for budget, scope, resources, and schedule are created in this phase. Remember, planning is not an exact science. If a project is given to two different teams, they may generate different plans.

Here are some keys to managing the planning phase:

- Use a project definition document

A common tendency among project professionals is to shortchange the planning process; instead, they jump right in and start working. This is a massive mistake. The time that is spent planning the project properly reduces time and cost while maximizing the quality over the life cycle of the project. Project definition is the key deliverable in this phase. It gives a clear description of all aspects of the project at a high level. Once the project has been approved by the client and other key stakeholders, it sets a baseline for the work to be done.

- Create a planning horizon

Once the project definition document has been prepared, it is time to establish a planning horizon by creating a workplan. A workplan provides stepwise instructions for project deliverables and project management. You can use a model workplan from a previous, similar project, or build one from scratch with a Work Breakdown Structure (WBS) and a network diagram. A detailed workplan needs to show how resources will be assigned, and should contain initial estimates. This sets your planning horizon. Past this horizon, there is a higher degree of uncertainty, which further planning should take into account. As the project progresses, the planning horizon advances as well. Activities that were initially too vague to plan in detail can be better defined as they approach the planning horizon.

- Create project management procedures

Project management procedures describe the resources available for managing the project. They define the protocols for managing risks, changing scope, and communicating. They also state the expected quality of work. If common procedures are already established and applicable, you can use them for your project. As a project controller, you should not only understand these procedures, but also ensure that all other team members and stakeholders do as well.

Execution Phase

This is the third phase of the life cycle. During this phase, the activities prescribed by the project management plan are performed. Proper control and communication strategies need to be implemented. The progress of the project is monitored continuously, and if adjustments are required, they should be recorded as variances from the original plan. Project professionals spend more time in this phase than in any other. During this phase, team members carry out tasks and report progress information to the manager in regular meetings. The manager and the controller use this information to keep the project on track by comparing progress reports with the original project plan. They measure the performance of the project activities, and take corrective or preventive action when required.

When progress reports deviate from the original plan, the ideal goal is to bring the project back to the original plan. However, if impossible, it is the responsibility of the project team to record changes from the original plan, as well as to publish all modifications. Sponsors and other key stakeholders need to be kept informed of the status of the project at an agreed-upon frequency and with an approved method of communication. The status reports must emphasize the anticipated end point in terms of cost, schedule, and quality of deliverables.

The project manager must perform number of core tasks during the execution phase. These include:

- Directing project objectives

Project managers must set clear objectives and realistic timelines from the very beginning of the project, because the majority of the budget will be spent during the execution phase. This means that before the project gets into full swing, the scheduling and phasing details related to the project tasks have to be set.

- Performing quality assurance

The project manager must carry out measurements and tests to determine if standards are being met, and should make adjustments if needed. The data from these tests should be shared with stakeholders to ensure that the project is operating according to the schedule and budget.

- Acquiring and managing the project team

Project managers must set up an efficient project team. Once the team is formed, the managers need to communicate expectations and timelines to team members. Team members should understand how the objectives of the project align with the overall organization's strategic goals. The team needs to have the right tools and knowledge to perform their tasks effectively, and building team cohesion and offering additional training materials bolster project success. It is important for the project manager to remain in tune with his or her team; this fosters a productive synergy between divisions of the project workforce. Project managers that listen and respond to their team's concerns and needs greatly increase the project's chances of success. When necessary changes are made and challenges are mitigated, quality products can be achieved.

- Distributing information

Routine communication is a detail often overlooked by less seasoned project managers. It is key to know whether to pick up a phone or to meet in person to discuss setbacks and schedule changes. During execution, an efficient communication network must be established and monitored by the project manager.

- Managing stakeholder expectations

Project managers must also monitor the expectations of the stakeholders, and communicate regularly with progress updates. They should take note of key benchmark objectives desired by the stakeholder, and respond efficiently to emerging concerns.

Managing and Controlling Phase

Bear in mind that no project proceeds exactly as planned. Good project professionals adapt to challenges by applying their management skills with discipline and rigor.

Keys to addressing challenges include:

- Reviewing the workplan regularly to measure progress in relation to schedule and budget. Based on project size, status reviews can take place weekly, bi-weekly, or at another appropriate frequency.

- Identifying completed activities, and updating the workplan accordingly. This way, you can easily pinpoint incomplete activities. Once you update the workplan, reassess whether the project can still be completed with the initial budget and schedule. If not, make any necessary adjustments.

- Constantly monitoring the budget. Assess expenditures by comparing the actual spending rate to the initially estimated rate. If the spending rate is too high, be proactive! Work closely with the team to determine a way to complete the work within the budget, or indicate a risk for anticipated budget overruns.

Be aware of red flags! Look out for signs that the project is in trouble. Warning signs include:

- Small variances in time and budget. These often appear at the very beginning of the project, and tend to grow bigger. Most people assume that they will overcome these variances, but this is a sign that they need to correct something. Be proactive; these variances may have damaging effects.

- Activities still underway that you thought were already complete. For instance, if you have people within the platform not yet moved to a new platform when they are supposed to already have migrated.

- Workers relying on unscheduled overtimes to meet deadlines. This often happens at the very start of the project.

- Drastically declining morale of personnel and team members.

- Deliverables and services deteriorating in quality. This could refer to the users beginning to complain that their converted emails are not working well and hence affecting communications within the project.

- Cutbacks in quality control procedures, testing, and project management time. If you are part of a big project, these factors have a tendency to affect everyone in the organization. Ensure that you are not cutting back processes that smooth the flow of project activities.

If these warning signs arise, turn to risk management procedures. Put together a plan to keep the project on track. However, in cases where you feel you cannot manage these situations, raise an issue!

So far, we have mainly discussed the basics of managing project budgets and schedules. Managing **scope** is equally important, and is a requirement of all projects. Many project failures occur not because of inaccurate estimates or lacking skill sets, but because of team members working on activities not originally included in the scope. Even if you have scope management procedures already in place, there are other ways in which scope can change. These include stakeholder demands and scope creeps.

- Stakeholder demands

In a large project, there can be many different customers, sponsors, and project workers. Requests for changes in scope typically arise from these stakeholders. Requests for different scope changes by different stakeholders may lead to friction. All changes have to be approved by the sponsors, since they are financially responsible.

- Scope creeps

Adding new functions or deliverables to the project are examples of changes in scope that most managers recognize. Unfortunately, many project managers fail to recognize that proper scope management must also be applied to even smaller changes. While one small change may not appear to have any effect, these changes accumulate, and eventually add significant cost and time to the project. **Scope creeps** are a series of small changes that are made to the scope without using proper scope change management procedures. Many projects fail as a result. Therefore, stay vigilant and diligent when managing all scope changes.

In addition to managing scope, while planning project activities, the project team is responsible for identifying any possible risks that may jeopardize the project. For each identified risk, the team must report the probability of it occurring, and describe its potential impact. Once this information is gathered, the risks should be organized according to priority. All events identified as "high risk" need to have specific prevention and mitigation plans. Additionally, all events deemed "medium risk" should be carefully evaluated and managed proactively. "Low risk" events are considered inconsequential because they are low probability events with minimal negative impact. It is important to note that some risks are often inherent, especially in a complex project. Perform risk assessments periodically throughout the course of a project; previously unidentified risks may surface unexpectedly.

You have to understand that issues are problems! For instance, consider an exchange migration program. If the exchange servers delay and are not delivered on time. That means that more time will also have to go into configuring them. To manage these issues, the project manager has to ensure that they are able to open these issues with diligence in order for them to be resolved fast. If there is no urgency attached to the resolution of such issues, the truth is that it may not really be an issue.

It may serve as a potential risk/problem. Real issues naturally have to be resolved with a high degree of urgency.

Closing Phase

In the last phase of the life cycle, each deliverable produced has to be measured against accepted criteria. Once all of the deliverables have been produced, and the customer has accepted the final product, the project is ready for closure. This phase involves providing deliverables to clients, delivering documentation to the organization, closing contracts with suppliers, releasing staff and equipment, and ensuring that the stakeholders are aware that the project is closed. When the project is terminated, its overall success is reported to the sponsors.

The project manager must list all activities to be carried out during the closing phase in a project closure report, which is created at the beginning of the phase. This ensures that the project is closed smoothly. Once the report is approved by the sponsors, the activities of closure detailed in the report are set in motion.

To evaluate the success of the project, the organization completes a post-implementation review one to three months after the project is closed. This is to document any benefits experienced after the project was completed, and to identify and record lessons learned during the project.

The project closure template suite is useful when generating project closure reports and post-implementation reviews. This provides you with a series of comprehensive document templates, which boost efficiency, and save time and effort.

CHAPTER TWO

Programs and Program Management

A **program** refers to a group of related product development projects that are managed in a coordinated manner. This allows the benefits of all the projects to be pooled, enabling stakeholders to access benefits that would have been unavailable if each project was managed individually. In the same way that successful projects need project managers, successful programs need program managers. **Program management** is the coordinated management of a group of related product development projects. In particular, it is geared towards coordinating projects governed by a contract between business enterprises. Customer contracts, proposal requests, and internal business plans play a central role in the program management process. Program management initiates other key product development processes, and establishes oversight of financial, technical, and scheduling goals.

One of the challenges for project and program managers is that it is almost impossible to disseminate and update large volumes of product data in an effective, timely manner. This is especially the case when complex products are developed within a complex environment. For instance, a product might be developed by multiple enterprises, consisting of dispersed teams, working on separate networks, in many time zones, and speaking multiple languages.

Compounding this data volume problem is the inherent sophistication of exchanging and monitoring product development information. It is hard to track individual progress, prevent work in isolation, and halt the use of out-of-date information, especially for those operating outside the main work stream. Many times, decisions that emerge from erroneous information lead to wasted time, money, and effort. Therefore, it is important to optimize all processes within the program management body. If the program management is well-defined and consistently executed, then teams can be kept in communication efficiently, and set targets can be met. Typically, benefits from improving the program management process include:

- ✓ More visible milestones, dependencies, and schedule changes
- ✓ Reduced time spent gathering performance metrics
- ✓ Improved performance
- ✓ Suppliers and partners brought into the program more efficiently
- ✓ Improved communication of accurate, up-to-date status information on cost, schedule, risk, and progress
- ✓ More effective collaboration
- ✓ Data shared more efficiently with suppliers, partners and distributed development teams
- ✓ Maximized value of in-person and virtual meetings

Program Life Cycle Phases

A program, just like a project, has a definite beginning and end. Like a project, a program has a life cycle. The program's processes are also divided into phases.

| Program Start-up phase | Program definition phase | Program Establishment phase | Program Management Phase | Program Closing Phase |

Figure 4. Program phases.

- ✓ Start-up Phase

In this phase, opportunities are evaluated, and decisions are made regarding whether or not to define the program. A program schedule is created. During this phase, an organizational plan is developed, detailing the allocation of work between internal teams, suppliers and other partners. Additionally, criteria is outlined for determining whether the program proceeds into the execution step.

- ✓ Definition Phase

The program is defined, documented, and evaluated, and a Program Definition Document (PDD) is created, as well as a program business case. Once these are approved, the program proceeds to the next phase.

- ✓ Establishment Phase

Resources, control, and infrastructural processes are acquired and implemented. Team members and stakeholders in the program are kept informed and trained in their roles and responsibilities.

- ✓ Management Phase

The projects that comprise the program are executed, and their deliverables and outputs are sent to the clients. During this phase, the execution of distributed development projects is monitored through progress reports, which raise issues, risks, and costs. Issues are resolved as they arise, and information flow is coordinated across dependent projects. Required program-level decisions are made in order to manage risk, facilitate communication, and ensure that cost targets are met. All activities consist of reviewing in-progress designs as well as validating customer requirements. Any changes in statements of work or in contracts have to be negotiated and managed. Additionally, any reports on contract deliverables, supplier deliverables, Earned Value, and technical performance measures must be prepared.

- ✓ Closing Phase

Once the program deliverables have been completed, the responsibility for operational control is handed over to line management. Certain measurements must be made in order to judge the success or failure of the program before closure. The OEM receives supplier notification of completion, and customer acceptance of final deliverables. All reports on project milestones are then submitted so that the program is transitioned to manufacturing, service, and support. However, if it becomes clear during the program that completion cannot come about successfully, the program may be terminated earlier.

The most important thing to note is that it is crucial for the program to begin well. It is a global truth that it is better to get things right the first time than to have to backtrack and fix them later. In other words, if the program is properly designed during the early phases, life during subsequent phases is considerably easier. The life cycle is designed to help the program managers establish their programs correctly.

Keys to preparation include:

- ✓ Providing a road map. This is particularly useful for teams new to program management, so that they understand the requirements and challenges that will be placed upon them.

- ✓ Specifying that relatively small amounts of work are commissioned and undertaken in the early stages of a program, and progressively scaled up as more information and certainty are achieved. In this manner, commitments of cost and effort are balanced against the levels of risk in the program.

After starting off on the right foot, it is important to maintain momentum throughout the project. Program execution offers guidance on how to avoid squandering early gains. Additionally, closing properly ensures that experiences and lessons learned are captured and made available for future programs. A program should be terminated in an orderly manner, with benefits documented and learning points captured and disseminate.

CHAPTER THREE

Portfolios and Portfolio Management

A **portfolio** is a collection of projects, programs, and other operational work. These are grouped together for more effective management, and to help achieve the strategic goals of the organization. It is important to note that not all projects and programs of a portfolio are interdependent or directly related.

Portfolio management refers to the centralized management of one or more portfolios. Management tasks for portfolios include identification, prioritization, authorization, project and program management, and related work. Unlike project management, portfolio management creates a forum for executives and senior management professionals. They collectively partner, agree, and act upon an optimum investment portfolio that offers strategic support for objectives and reconciles competing interests within the organization.

Always bear in mind that portfolio management is an ongoing endeavor, and can be updated from time to time at the discretion of the organization.

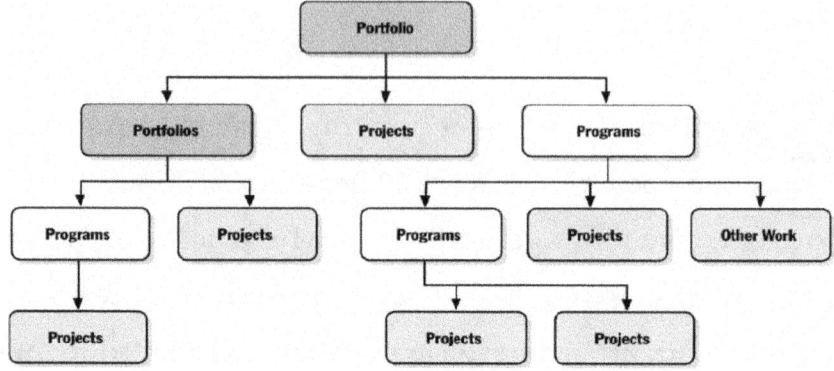

Figure 5. Relationship between portfolios, programs, and projects.

Portfolio Processes

Portfolio management involves a set of related continuous business processes. These processes assist in making decisions regarding the portfolio's content, and are carried out by a portfolio manager or a specific management team. The main foundation for portfolio processes is a comprehensive database of information on project-oriented initiatives, which stores data on the finances, organization, and performance of key projects. Both managers and executives can explore and refer to this database.

Portfolio processes play a crucial role in informing project justifications, allocation of resources, and prioritization. They also help maintain the visibility of key project information across the enterprise. Portfolio processes are collaborative; they rely on shared information to arrive at a consensus regarding how projects align with the overall strategic goals of the organization. They also allow for quick access and communication of relevant information both horizontally and vertically within the organization. This ensures that the decisions made about projects are focused, prioritized, and efficiently drive resource allocation, use, and future investing.

Portfolio processes are divided into groups. Groups and their individual constituent processes are often put through several iterations. The constituent processes also have the ability to interact. Interactions can occur not only between processes of the same group, but also between processes of other groups. The two groups of portfolio processes are the Aligning Process Group and the Monitoring and Controlling Process Group.

- ✓ **Aligning Process Group**—these processes play a role in determining how portfolio components will be categorized, evaluated, selected and managed.
- ✓ **Monitoring and Controlling Process Group**—these processes involve reviewing performance indicators to assess if portfolio objectives align properly with the overall strategic goals of the organization.

The Aligning Process Group ensures the availability of current information regarding the strategic business goals of the portfolio. It also ensures the availability of current rules for evaluating components and managing the portfolio. Moreover, this process group plays a significant role in establishing a well-structured, agreed-upon technique for keeping portfolio components aligned with the organization's overall strategic goals. Therefore, the Aligning Process Group is very active when the organization reviews and updates its strategic goals, and is used to define short-term budgets and plans for the entire organizational body. Traditionally, these activities took place annually, but some organizations now have more frequent budgeting cycles. They may occur quarterly, for instance, or be scheduled according to changes in the business climate.

Alternatively, the Monitoring and Controlling Process Group is more concerned with ensuring that the portfolio is performing according to predefined metrics set by the organization. Some metrics, such as Return on Investment (ROI) and Net Present Value (NPV) thresholds, are used to monitor the aggregate performance of all portfolio components. Other metrics are used to track individual components of the portfolio.

There are nine portfolio processes: identification, categorization, evaluation, selection, prioritization, portfolio balancing, portfolio authorization, portfolio reporting and review, and strategic change.

1. Identification

This process creates and evaluates an up-to-date list of ongoing components (projects, programs) and newly proposed requests/demands. The first task of the identification process is to compare ongoing projects, programs to new proposals, using predetermined definitions and key related descriptors. The second task is to reject requests/demands that do not fit the definition. The third and final task is to put the new, approved requests/demands into predefined classes. Such classes include: project, program, and other works.

2. Categorization

During this process, classes and their components are placed into relevant business groups. These business groups serve as important categories to which common decision filters can be applied for evaluation, selection, prioritization, and balancing. Key tasks of categorization include identifying strategic categories based on the strategic plan, comparing components and identifying potential categories for them based on categorization criteria.

3. Evaluation

In this process, pertinent evaluation factors are developed and used to assess each component. Key tasks conducted during this process include evaluating components with a scoring model comprised of weighted key criteria, producing graphical representations to facilitate decision-making in the selection process, and making recommendations for the selection process.

4. Selection

During this process, evaluated components are formally rejected or approved for further consideration. The key activities of this process are comparing the evaluation results of the components, and approving worthy components according to selection criteria.

5. Prioritization

This process ranks approved components within each strategic or funding category based on previously established criteria. Key activities of this process include confirming that each component fits within its strategic category, and ranking components according to scoring or weighting criteria, which determines which portfolio components receive highest priority.

6. Portfolio Balancing

In this process, the portfolio component mix is created. The aim is to create a portfolio with the greatest potential to support the organization's strategic initiatives and objectives. Key activities of this process are adding components that have been selected and prioritized, identifying unauthorized components, and marking components for suspension, reprioritization, or termination.

7. Portfolio Authorization

Formal authorization communicates to stakeholders that the portfolio is properly balanced. Key activities of this process are announcing portfolio-balancing decisions, re-allocating the budgets and resources of inactive and terminated components, officially allocating sufficient financial and human resources to authorized components, and communicating the expected outcomes in terms of timeline, performance metrics, and required deliverables for each authorized component.

7. Portfolio Reporting and Review

During this process, performance indicators are gathered, and periodic reports are generated. The portfolio is assessed routinely for proper alignment with organizational goals. Other key activities include conducting reviews for component sponsorship, accountability, and other ownership criteria, using organizational governance standards as a reference. Additionally, component priorities, dependencies, scope, expected return, risks, and financial performance are weighed against portfolio control criteria and organizational perceived value and investment criteria.

The reporting and review process also provides a thorough assessment of how business forecasts, resources, and capacity constraints are expected to impact portfolio performance. This helps determine whether to continue with, add to, or terminate specific components, and whether to reprioritize and realign them with strategic goals. Other major activities are making recommendations and proposing changes to component management.

8. Strategic Change

This is one of the most important processes. The strategic change process ensures that portfolio management responds to changes in strategy. Unlike the other processes we discussed, this process involves countless activities that are as widely varied as the organizations using portfolio management.

Portfolio Life Cycle Phases

Like programs and projects, portfolios have life cycles with distinct phases. Portfolio phases include inventory, analysis, alignment, execution, and closing.

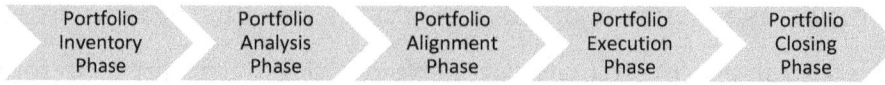

Figure 6. Portfolio phases.

Inventory Phase

The inventory phase defines the portfolio and offers support to the analysis phase. Information is obtained from valid sources, which may include spreadsheets, project management systems, accounting systems, and interviews of both project participants and project managers. To start, collect key information from ongoing projects. Next, identify the priority process or gating process in order to make decision on what you would like it to be. Then, determine the category under which each component falls, and what strategic initiatives the organization has.

Analysis Phase

In this phase, the strengths and weaknesses of the portfolio are assessed. What is being done well? What isn't being done well? In this phase, you can perform countless types of analysis. These include a basic project justification, which simply identifies component projects that are facing similar problems. Some projects could perhaps be combined, and others may even be cancelled altogether. A sort of resource categories may reveal potential shortcomings; if shortcomings are detected early enough, you can thoroughly consider such options as bringing in contractors, making cancellations, or increasing headcounts. Sorting by department may reveal such important aspects as customer service. Once analysis is complete, it is time for managers and executives to assess different dimensions of the portfolio, in order to get an in-depth understanding of the entire model.

Alignment Phase

The alignment phase often results in a changed portfolio. Each component in the portfolio becomes a candidate for reclassification; projects previously on hold may enter an active state, or may even go from an active state to cancellation. A project that started as just an idea may become a fully feasible option. In this phase, it is crucial to commit resources to creating new ideas for potential projects, in case the current portfolio does not adequately address certain strategic concerns.

Deciding which projects to cancel or delay is very difficult. To facilitate these decisions, alternative project portfolios are evaluated using "what if" scenarios. It is critical to note that the key challenge in this phase is avoiding the temptation to modify decision-making criteria, or even the project data itself, in order to make the present ongoing portfolio seem optimal.

Management Phase

In this phase, the management body is given full access to the portfolio and is able to collaborate on issues relating to component prioritization, budgeting, categorization, scarcity of resources, investing options, milestones, and delays. Indeed, this is the phase where true portfolio management takes place. In other words, in this phase, managers make sure that projects are aligned with the organization's overall strategic goals. They cut projects, and redistribute budgets and resources. However, a challenge is how to communicate portfolio information as actionable plan. Here, automation can go a long way in sharing information on updated project priorities, budgets, schedules, and resources to relevant departments and managers.

A major step in this phase is mobilization. Mobilization requires that information, tools, and resources are given to department and project managers in a format that best suits their needs. This is the phase in which project professionals implement the portfolio processes and their best practices.

Closing Phase

Just like in a project or program closing phase, portfolio closing involves following up and ensuring that all portfolio objectives were met. During this phase, all key stakeholders must accept closing activity checklists by signing off.

Valuable lessons learned during the implementation of portfolio projects must be documented properly for future reference. The portfolio objectives have to be assessed to ensure that there are no loose ends. This phase also opens up opportunities to follow up with contractors that supported the team in achieving their deliverables.

In this phase, key steps have to be followed. These steps include:

- Analyzing portfolio project performances to ensure that the set goals were achieved, and that any problems raised were addressed using an already prepared checklist

- Analyzing team member performances to ensure that everyone met their goals along the set timelines without compromising the quality of work

- Documenting project closure to ensure that all aspects of the portfolio projects are completed, and that reports are compiled and shared with the stakeholders

- Conducting post-implementation reviews of the entire portfolio and taking into account any lessons learned

- Accounting for all used and unused budgets, and ensuring that any resources that remain are allocated to future work

CHAPTER FOUR

Project Stakeholders

Types of Stakeholders

Stakeholders include all the individuals and organizations who are actively involved in the project. In other words, people who may affect or be affected by the activities, decisions, and outcomes of the project are all considered stakeholders. Stakeholders include not only people whose interests are affected positively, but also those who are affected negatively. For instance, a local environment group can be described as a negative stakeholder for a refinery project, as their mandate is to protect against any pollution of the environment.

There are three broad categories of stakeholders:

1. Stakeholders within the project, such as the project team and project leaders

2. Those outside the project, but within the organization, such as sponsors, company employees, and other organizational groups

3. Those that are outside the organization entirely, such as sellers, partners, suppliers, customers, civilians, and government bodies

It is the duty of project professionals to identify all stakeholders and their requirements, and to then ensure that they meet those requirements. It is important for the project controller to understand that the process of identifying stakeholders can often be difficult.

The following diagram provides a visual of the three groups of stakeholders. The center circle represents stakeholders that are within the project and are working on project activities. The next three circles represent stakeholders who are outside the project, but lie within the organization. The outer two circles represent stakeholders outside the organization.

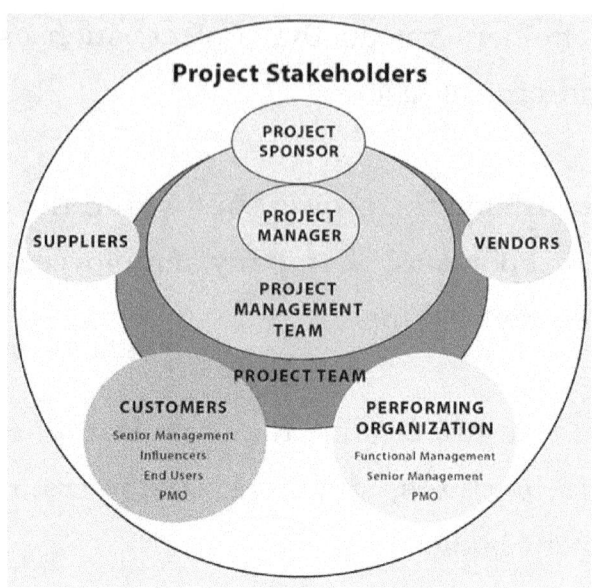

Figure 7. Categories of stakeholders.

Key stakeholders within the project include:

✓ Top level managers

The **project manager** is at the top level of project management, and is responsible for directing the project and coordinating the work of the project team. There are other top level managers, including company directors, presidents and vice presidents, division managers, and corporate operation committees. It is the role of these people to direct the strategy and development of the organization.

A project manager's role is to recruit the best staff to conduct project work. It is also important for him or her to acquire all necessary materials and resources, and to be visible, enhancing his or her professional standing in the company.

Typically, it is the role of the company director to choose assignments and decide who works under the project manager. It is important for project managers to keep their superiors informed, in order to ensure that they receive the necessary resources to complete their projects. In case anything goes wrong, it is helpful to have an understanding and supportive boss.

- ✓ Project controller

The **project controller** has a role in advising the project team and informing the project manager of potential cost and scheduling issues, as well as in coming up with effective recovery plans. A potential downside of this role is its visibility. In other words, as project controller, your face is known. This means that if a large project fails, the cost of failure is quite substantial for you personally.

Therefore, to ensure that you drive the project towards success, you need to develop in-depth plans and identify major milestones, which are to be approved during the planning and design phases. Additionally, it is important to actively seek out your top level managers for any information needs. As project controller, you must develop a methodology for distributing status reports on a scheduled basis, and stay informed of project risks and potential impacts at all times.

✓ Peers

Peers are a group of people who are at the same level in the organization as the project leaders, but were not selected to lead the project team. In other words, they may have a vested interest in the project and product, but lack the leadership responsibilities and accountability to control the project. As a project controller, interactions with your peers can be impeded by a number of factors:

- Inadequate control over peers
- Political maneuvering or sabotage
- Personality conflicts and technical conflicts

- Envy, because your peer may have wanted to lead the project
- Conflicting instructions from your manager and your peer's manager

In order for you to work smoothly with your peers, you have to ensure that you have their support. Most of your peers serve their own self-interest; hence, you have to use your best selling, influencing, and politicking skills here. Additionally, ensure that the sponsors of the project offer you support, so that you are empowered and given as much authority as possible. In other words, the sponsor can make it clear to your peers that their cooperation on project activities is expected. This gives you the authority to confront a peer if you notice a dysfunctional behavior that seems out of line with your established goals and standards of performance.

✓ Team members

These are the people involved in performing the activities of the project. A team is composed of people who are dedicated to the project full-time, and also people who work on the project on a part-time basis. Team members rely on the leadership of their project manager for direction and support. When people work closely as a team, they build a rapport, which gives them a better chance of solving and learning from problems as they arise.

When dealing with project team members, take note of some possible challenges:

- Members having priorities other than the project. They may be juggling many projects as well as their full-time job, and may have difficulties meeting deadlines
- Personality conflicts due to differences in work style and social values
- Members not communicating that they have missed deadlines until it is too late to recover

The project manager must possess strong interpersonal skills in order to manage the project team effectively. He or she should involve the members when planning the project, arrange private and informal meetings regularly, be available at all times to hear concerns, motivate the team members, and complete team performance reviews in a timely manner. Team members must seek out proper support from the manager, so that the manager will in turn support them. The best ways to ensure proper support are:

- To understand exactly how performance will be measured
- When unclear about directions, to ask for clarification
- To develop a reporting schedule that is acceptable to the manager
- To communicate frequently

Key stakeholders outside the project sphere include:

✓ Customers

Customers, also referred to as **clients**, are organizations or individuals that will receive the final product of the project. A customer can be a single entity, or have multiple layers. It is important to note that sometimes the customer both purchases and consumes the product; other times, the customer purchases the product for further distribution to **users**, the final recipients who directly consume the product.

Customers can be classified as internal or external.

- **Internal customers**

This refers to a group of individuals within the organization who are customers of projects that fulfill internal demands of the organization. Internal customers hold the power to accept or reject project products. During the early phases of the project, it is the role of the project manager to negotiate, clarify, and document project specifications and deliverables. Once the project starts, he or she must stay tuned in to concerns and issues raised by the customer, while ensuring that customers are informed every step of the way. All project professionals must understand the needs of the customers, clarify key project requirements, specify the procedure for changes, and establish the central point of communication.

There can be a number of stumbling blocks when dealing with internal customers:

- Lack of clarity about precise product specifications
- Lack of documentation
- Lack of knowledge of the customer's organization and operating characteristics

- Unrealistic deadlines, budgets, or specifications requested by the customer
- The customer hesitating to sign off on the project or accept responsibility for decisions
- Project scope changes

External customers

External customers are a group of people outside the organization to whom the project is marketed. For instance, external customers are the buyers of automobiles from an automobile company.

✓ Sponsors

Sponsors are a group of individuals who are responsible for the financial resources for the project.

To recap, the main stakeholders in a project are top level project managers, project controllers, peers of project leaders, team members, customers (internal and external), and sponsors. However, it is important to note that there may be additional stakeholders involved in a project that may either be internal or external to the organization. These stakeholders can be owners, funders, team members, families, government agencies, sellers and contractors, citizens, and media outlets, among others. Classification of all these groups is very important in determining who the stakeholders are, and what their stake is in the project. In some cases, the roles and responsibilities of different stakeholders in the project can overlap. Managing all the stakeholders' expectations and interests can be a very difficult task, because their differing objectives can conflict.

Organizational Influences

Often, projects are part of a larger project of the organization. Even if the project involves a partnership or joint venture with other institutions, it is important to note that the project will be principally influenced by its own parent organization. The level of maturity in the organization (culture, project management system, Project Management Office (PMO), structure, and style, among other features) can play a big role in influencing the project.

For instance, the culture of the organization often directly impacts the project. Most organizations develop a unique and tangible work culture. It is through the organization's shared values, norms, and beliefs that its culture is born. Culture is also reflected in procedures, policies, and relationships with authorities.

Project-based organizations are organizations whose operations are typically projects. These organizations can be classified into two major categories:
Those that derive their revenue by performing projects for others; and those that have adopted **management by projects** as a system for all their internal operations. Organizations in the second category often have their entire management system set up to oversee projects. For instance, a financial system can be set up to provide management of accounts, tracking, and reporting for several projects simultaneously.

On the other hand, **non-project-based organizations** lack this coordinated project management system. A coordinated system is extremely important during project implementation and control, and helps ensure that project staff are actively working to achieve the objectives of the project. The absence of this system makes project management difficult. In some cases, non-project-based organizations adopt departments or sub-units that operate as sub-organizations with project-based systems.

The organization's structure impacts the project because structure influences the manner in which resources are made available to projects. Structures span a spectrum from **functional** to **projectized**, with a wide range in between. The classic functional organization structure refers to one in which each employee has one clear supervisor. In this case, employees are often classified based on their areas of specialty. Such organizations still run projects; however, the scopes of the projects are limited by the boundaries of their functions. When a new project is developed in a purely functional organization, it is referred to as a **design project** during its design phase. Whenever questions arise during this phase, they are directed to the department above it, which then in turn consults the next highest department.

On the other hand, a projectized organization often aggregates its members into departmental units that all fall under the authority of the project controller. This means that most of the resources in the organization are channeled towards performing the activities of the project.

Matrix organizations, alternatively, are a blend of functional and projectized organizations. Weak matrices often maintain several features of a functional organization; in other words, they have more traits of a functional organization than of a projectized one. The project controller, in this case, has more roles than just that of a coordinator. Alternatively, strong matrices have more features of a projectized organization. These features include a full time project manager with considerable control over the project; as well as a full time administrative project staff. Both managers and controllers are part of a centralized project support unit called the **Project Management Office** (PMO), or Project Office. The PMO has a wide range of uses in matrix organizations, but mainly offers support services to project professionals in the form of software, trainings, and templates. The results of the project are the responsibility of the PMO.

CHAPTER FIVE

The Project Controller

Emergence of the Project Controller

The emerging position of the project controller has changed the project management landscape in profound ways. Initially, a "coordinator" primarily handled administrative chores, such as data entry, collection of status information, and the design of status charts. However, organizations still had needs regarding the development of estimates and schedules, which led to the creation of a "planner" position. The majority of firms used this role to track risks, keep issue logs, organize status report sessions, analyze schedules, and facilitate all types of planning. In recent times, there has been a surge in the planner's responsibilities. These include financial reporting of Earned Value, handling resource allocations and constraints, creating schedules and critical-path analyses, and providing other documentation needed to comply with regulatory requirements.

Recently, the roles of planner, coordinator, and scheduler have morphed into that of a single project controller. The project controller supports the project manager by handling most of the detail-oriented, time intensive, and analytically focused project tasks. Consequently, the project manager is free to focus on strategic project goals and objectives. In many cases, the manager is even able to take on additional projects. Best practiced organizations look at the project manager as the CEO of the project, and the project controller as the CFO; these two roles operate independently of each other, but are nevertheless critical to each other's success.

To further explore this analogy, we will define the relationship between the CEO and CFO. The CEO develops the overall vision and sets the stage for leadership of the entire organization. On the other hand, the CFO ensures the financial viability of the firm by following strict industry policies and procedures. Both the CEO and CFO work collaboratively to hit organizational targets. Similarly, both the project manager and project controller carry out crucial duties. Both possess significant skill sets and have responsibilities necessary for establishing projects in a timely fashion and within estimated budget limits.

Key General Project Controller Skills

When it comes to project management, the project controller is charged with monitoring all aspects of an ongoing project. He or she oversees a wide range of activities:

- ✓ finances and accounting
- ✓ research and development
- ✓ strategic, operational, and tactical planning
- ✓ organizational structure
- ✓ personnel supervision
- ✓ benefits and career paths
- ✓ management of work relationships through trainings
- ✓ motivation
- ✓ delegation
- ✓ conflict resolution

Project controllers must also manage themselves! They must employ techniques for time and stress management. General management skills allow a project controller to build up their project management skills. Five important general management skills include:

1. Leadership

According to Kotter, there is a distinct difference between leading and managing. In spite of this difference, a successful project controller has to both lead and manage. When it comes to managing, the key goal is to produce good results consistently. Leading, on the other hand, is about establishing direction, getting people on board, and offering motivation and inspiration. A project controller is expected to contribute to the leadership of a project, especially for larger projects. However, this expectation is not limited to just the project controller, as different individuals at different times may be required to step up and demonstrate leadership during the course of a project. Leadership has to be demonstrated at all levels of the project. These levels include project leadership, team leadership, and technical leadership.

2. Communication

Communication is the effective exchange of information. The person that is sending information is responsible for ensuring that the information is clear, complete, and free of any ambiguity, to ensure that the receiver is getting the right message. On the other hand, it is the responsibility of the receiver to ensure that the information received is complete and well understood. Communication can be spoken or written, internal or external (between team members and managers, or via media outlets, customers, and the general public), and vertical or horizontal (up and down the organizational hierarchy, or between peers and partner institutions).

Communication often involves a substantial body of knowledge unique to the context of each project. The context of the project determines several communication options, including how the sender-receiver model is established (such as feedback loops to avoid communication barriers), the choice of media, writing styles (passive versus active voice, structure of sentences, and word choice), presentation techniques (visual aids, audio, designs, and body language), and how management agendas are addressed.

3. Negotiation

This involves the ability to persuade others to compromise or come to terms with you. In other words, it is all about coming to an agreement. In order to arrive at an agreement, there have to be direct negotiations, which can be assisted by mediations and arbitrations. Negotiations can occur around different issues, during different times, and at varying levels of the project. Often, the project team may have to negotiate many of the following factors during the course of a project: scope, cost, budget schedules, objectives and their respective schedules, terms and conditions of the contract, assignment, resource allocation, and changes to any of the above.

4. Problem solving

Problem solving involves a combination of problem definition and decision making. In problem definition, one has to distinguish between the cause of a problem and its symptoms. A problem can be internal or external to the project. Problems can be technical, managerial, or interpersonal. On the other hand, decision making involves identifying possible solutions and choosing between them. It is important for a project controller to understand that once a decision has been made, it has to be implemented. Decisions are also affected by timing. That means that even if you have the "right" decision, it may not be the best one if it is made too early or too late.

5. Influencing the organization

A successful project controller has to be able to get things done. This means that he or she has to understand both the formal and informal structures of the organization and of other stakeholder institutions. In order to influence the organization, the project controller must understand the mechanics of politics and power, as both play an important role in projects, and may have either a positive or negative impact. **Power** refers to a potential to influence behavior, change the direction of activities, overcome resistance, and persuade people to do things they might not normally do. **Politics**, on the other hand, refers to obtaining a collective action from a group of people who have varied interests. A project controller has to be able to use conflict in a creative manner. Failure to do this may result in power struggles and other negative political games that yield unproductive results.

So, beyond these skills, what makes a good project controller? To carry out his or her core responsibilities efficiently, a successful project controller needs to possess technical expertise in project management software, and in related spreadsheet and financial database tools.

The controller will also need business process expertise in risk analysis, cost budgeting and estimating, resource forecasting, critical-path diagramming and analysis, and change control. It is crucial to understand that the project controller might be involved with several simultaneous projects. This is often determined by the sizes of the projects and the experience of the controller. To handle multiple projects, he or she must continue to meet the required standards demanded by the organization, and exhibit the requisite flexibility for dealing with multiple project managers.

We will now consider a case study. The following figures will help you visualize all the skills that we have discussed, and will help you understand how to incorporate them into your project control role.

OB AND YOUR CAREER — Become a Level 5 Leader

Jim Collins, author of *Good to Great*, and his team selected 11 companies (Abbott Labs, Circuit City, Fannie Mae, Gillette, Kimberly-Clark, Kroger, Nucor, Phillip Morris, Pitney Bowes, Walgreens, and Wells Fargo) from more than 1,400 that had been listed in the *Fortune* 500 at some point or other. Each of the selected companies had mediocre results for 15 years and then went through a transition. From that point, they outperformed the market by at least three to one—and sustained that performance for at least 15 years. Each of these was compared with companies in the same industry and about the same size.

Using hundreds of interviews, Collins identified key factors that enable companies to move from mediocre institutions to great institutions. The comparison companies lacked these factors and failed to become great. Perhaps the most important component of the transition from good to great is what Collins designated as *Level 5 Leadership*.

Level 1 is a highly capable individual who "makes productive contributions through talent, knowledge, skills and good work habits."

Level 2 is a contributing team member who "contributes individual capabilities to the achievement of group objectives and works effectively with others in a group setting."

Level 3 is the competent manager who "organizes people and resources toward the effective and efficient pursuit of predetermined objectives."

Level 4 is an effective leader who "catalyzes commitment to and vigorous pursuit of a clear and compelling vision, stimulating higher performance standards."

Level 5 is the executive who "builds enduring greatness through a paradoxical blend of personal humility and professional will."

Every one of the good-to-great companies had Level 5 leaders in the critical transition phase. None of the comparison companies did. These leaders are described as being timid and ferocious, shy and fearless, and modest with a fierce, unwavering commitment to high standards.

Level 5 leaders rely on instilling inspired standards and not inspiring charisma to motivate. They build a culture of discipline. He or she is not a tyrannical disciplinarian, but one who enables freedom and responsibility. Self-disciplined people are hired who are willing to go to lengths to fulfill their responsibilities. They consistently adhere to what Collins calls the Hedgehog Concept, the intersection of three circles:

1. Brutally and realistically determining at what the company can be the best in the world and pursuing it in light of the next two points.
2. Deciding the most effective way of generating sustained cash flow and profitability, then determining the single most important indicator. In the case of Walgreens, it was profit per customer visit and not the traditional profit per store.
3. The good-to-great company and its employees do only the things they are deeply passionate about. This passion is not stimulated or imposed but discovered.

Level 5 leaders channel their ego needs away from themselves and toward building a great company or organization. They often will sacrifice their own gain for the gain of the company. When things do not go well, Level 5 leaders take responsibility for the failures and never blame other people, external factors, or bad luck. Level 5 leaders can help identify and develop other potential leaders throughout the organization.

Good-to-great companies set out on a path to improve long-term results that go unnoticed by the outside for years. They then suddenly appear, well on their way to becoming great.

All of the good-to-great leaders created standards and doggedly kept to those standards for the years of their tenure.

Sources: Adapted from J. Collins, "Celebrity Leadership," *Leadership Excellence* 25, no. 1 (January 2008): 20; Des Dearlove and Stuart Craner, "Jim Collins and Level 5 Leadership," accessed at www.management-issues.com/2006/5/24/mentors/jim-collins-and-level5-leadership.asp on May 16 and 20, 2007; and Jim Collins, *Good to Great* (New York: Harper Collins, 2001).

Figure 8. A case study illustrating exemplary key project controller skills.

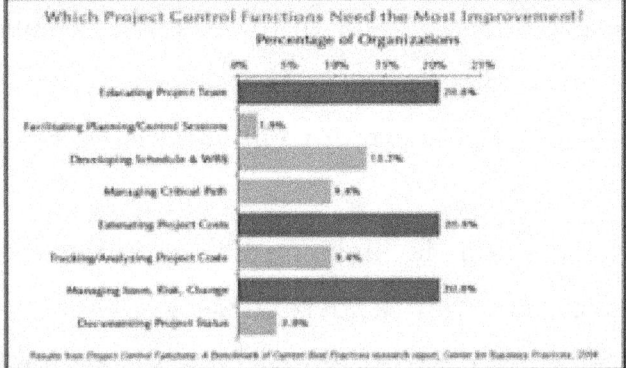

Figure 9. Illustrative statistics of how well organizations perform project control functions.

Project Controller Responsibilities

The project controller plays a prominent role within the project team. The project controller works to define the project's objectives and goals, and makes certain that there is proper planning and monitoring. The controller helps analyze progress reports based on work schedules. He or she creates and maintains the schedule and budget, and is also charged with the responsibility to recommend and initiate actions to improve progress.

As mentioned previously, the majority of organizations position the project manager and project controller as part of a centralized project support structure, which is usually called the Project Management Office (PMO), or the Project Office.

This approach is designed to enable consistently accurate documentation and reports. Usually, the PMO is directly under a high level business executive whose primary role is to drive the business in his or her domain. In the more regulated modern business world, these executives demand reliable and consistent status reporting across all their projects.

The project controller is a key contributor within the PMO. He or she bears the responsibility of carrying out functions that provide critical input to the reporting process. Some of these functions include:

- ✓ Educating the team on proper processes

The controller must make sure the team understands the basics of project management methodologies, policies, standards, and processes across all phases of the project (initiating, planning, controlling, executing, and closing). This includes educating them on scheduling and costing tools, describing the processes' purposes, detailing approval procedures, and determining who is responsible for deliverables.

- ✓ Overseeing and facilitating the planning and control of sessions

Sessions may deal with a variety of topics including estimating resource hours and task durations, managing risk, developing the Work Breakdown Structure (WBS), planning, developing the network diagram, integrating the WBS and network diagram into the project schedule, capturing lessons learned, and reporting.

✓ Developing the project schedule

This is also a core responsibility for a controller. Using organizational project office standards, the project controller works with other key team members to develop the project schedule, which includes resource leveling and critical-path analysis. Once set, the schedule is used to manage resource assignments, measure work progress, track milestones, and monitor and report on performance metrics.

✓ Monitoring progress

The project controller proactively monitors the schedule with the aim of delivering the project on time. As part of this responsibility, the project controller must:

- Analyze any variation or forecasted variation to the plan
- Regularly collect task statuses and update the schedule
- Emphasize items such as critical-path, slipped, and upcoming tasks
- Calculate, analyze, and present metrics such as Earned Value

- Manage resource assignments, which includes submitting resource requests
- Monitor current resource allocations, and forecast future needs
- Meet with the project manager to discuss project status

✓ Tracking and analyzing costs

The project controller must work with available project and accounting management software tools to track actual (or blended) costs, including equipment usage, labor hours, and procured resources. He or she must also analyze cost run rates and variations to extrapolate and forecast total costs, as well as report findings to the Project Office, project manager, and any financial oversight committees.

✓ Managing core issues, risks, and internal and external changes

The project controller ensures that emergent issues, potential risks, and change requests are identified, analyzed, and estimated. Metrics must be delivered to the appropriate management levels for disposition, and also documented in the project's collaboration database.

✓ Documenting and delivering status information

With many organizations applying new corporate governance standards, the project controller must now be responsible for developing Earned Value reports, and for presenting material to executive management. Although the project manager is tasked with corporate communications, it is the project controller who performs the analysis and forecast, and who makes recommendations.

Types of Project Controller

A difference between the role of the project manager and that of the project controller is that the controller's responsibilities evolve based on the project's context. Three types of project controller, also known as Project Controller Officer (PCO), exist:
1. The project PCO
2. The program PCO
3. The portfolio PCO

1. The Project PCO

The **project PCO** works with a project manager. The project PCO is responsible for plan control, budget control (invoice tracking), risk control, documentation, and some administrative tasks.

The project PCO job is usually the entry point to the project management world. However, with experience, the project PCO can become a program PCO or a project manager. Qualified project PCOs are organized, proactive, and have strong communication and organizational skills. It is important to be efficient, knowledgeable, and independent; make sure to understand requests so you can deliver appropriate output. Knowledge of tools such as Microsoft Project and understanding organization policies and processes are crucial for helping the project manager and team organize the work.

Be an expert in project management tools! Get training or certification in MS project or Primavera. The Certified Associate in Project Management® and PMI Certification are other assets that can help you acquire sufficient knowledge.

2. The Program PCO

A **program PCO** works with multiple project managers and program managers. This PCO assesses **Key Performance Indicators** (KPIs) and reports on the overall program. He or she facilitates the work of the project manager, program manager, project PCO, and project team by focusing on KPI creation and documentation of standard procedures that align with the company's policies and processes. Additionally, he or she helps different project managers with their project plans, but is usually less involved in the details. He or she is responsible for reminding managers of tasks and making sure that deadlines are respected.

To be successful in this role, a program PCO has to be proactive, study the team, and apply the appropriate strategy to obtain the right information on time. A proactive PCO can quickly manage certain vulnerable areas of the project. Organization and communication are crucial. Training in project management tools is essential in providing the appropriate support. This role involves skill in coordination.

3. The Portfolio PCO

The **portfolio PCO** primarily works with the portfolio manager or the Project Management Office (PMO) director. Furthermore, the portfolio PCO also works with the entire organization, project manager, program manager, sponsors, and vice president. This role is more strategic in nature, and is oriented around the control and respect of PMO processes. Sometimes, portfolio PCOs have to create and document PMO processes, the implementation of which may require changes in management strategy. To do this, they must communicate and collaborate with key people in order to create the process, explain benefits, perform retrospective reviews, document lessons learned, and focus on continuous improvement. In other words, PMO processes are not static; they evolve and change, and the portfolio PCO is highly involved in their control. Additionally, a crucial part of the portfolio PCO role is developing a model to plan the overall budget and resource capacity.

Training (certification in PMI's Portfolio Management Professional (PfMP)), knowledge of project selection, and prioritization strategy are important. At this level, the portfolio PCO has to understand the company's vision and strategic plan, as well as the projects' relationships and dependencies.

Why Organizations Need Project Controllers

Primarily, executives need to understand the ultimate impact and value of project management on their businesses. They may be suspicious of any new expenses introduced by the project, especially if those expenses involve expanding their staff to include a new project controller role. Educating corporate decision makers about the responsibilities of the new position is the most important facet of building a successful business case for the project controller role. It is crucial to emphasize the positive impact that it will have on the company.

Here are a few key points to include when educating corporate officials:

- ✓ There is a pressing need to strengthen communication with high-level executives. With the support of a project controller, project managers can share vital information more regularly, which includes financial reporting of Earned Value, and reporting other documentation needed to comply with regulatory requirements. As a result, executives will have a better understanding and more control of the projects and their financial implications. Additionally, decisions will be made in a more timely fashion.

✓ Through the project controller role, firms can leverage and expand the capacity of project managers to take on and effectively complete more projects concurrently. This way, the organization can efficiently derive more benefits than before. Moreover, the project manager would have the capability to effectively spend more quality time with executives in explaining statuses, gaining insight into political and budgetary issues, and selling project ideas.

✓ A project controller can help reduce the workload of the project manager, making the project more efficient. Many project managers face challenges dealing with the detailed tasks of project planning, and in many cases, prefer to handle high-risk issues, deal with executives, communicate statuses to the organization, and work on other high-visibility activities. By delegating important, yet time consuming tasks to a project controller, managers are free to focus on higher level issues. For instance, a controller can focus on boosting staff motivation, which plays a pivotal role in project management. They can also help avoid scheduling issues and communication failures. This means that more projects will be completed on time and within budget, saving the company from the debilitating cost of project delays and failures.

Role of a Project Controller in Budget Management

A budget has everything to do with success! Of course, very few companies have unlimited project budgets. Therefore, project stakeholders use a budget assessment as a litmus test for defining the success or failure of a project. Analyzing whether the budget was adhered to is a way to judge if the project controller has been at the top of his or her game in ensuring that all project activities met set budget limits. It is this fact that fuels the pressure on the project controller and the rest of the project team each day.

Here are five strategies that a project controller can apply to ensure effective and efficient control of project budgets, and to avoid succumbing to cost overruns:

1. Properly understand the stakeholder's needs and wants

It is the role of the project controller to ensure that the project team and sponsors have a solid grasp of the true desires of the stakeholders. This makes it easier to pinpoint key project requirements. As a project controller, you have to seek out a deeper understanding of what stakeholders in the project expect. It is their expectations regarding deliverables and other requirements that play a vital role in defining the budget from the start of the project.

This is not as simple as it sounds. Often, the needs and expectations of stakeholders in a project are never properly identified. Many projects are initiated around the stakeholders' basic needs, and then later expanded to include additional wants, which put the project at a high risk of overrunning the budget. This leads to great disappointment for everyone, and does not reflect well on the organization, jeopardizing future sponsorship opportunities and partnerships. Therefore, the very first steps for managing the project budget are accurately defining, documenting, and curating all project requirements put forth by the stakeholders. These requirements are then communicated to all partners and members involved in the project, either directly or indirectly. It is very important to perform this step before budgets are set.

2. Budget for surprises

Something that many project controllers overlook is the importance of setting realistic cost estimates. It is important to plan for surprises, so that you are not blindsided. All team members and applicable stakeholders should give their input regarding possible expenses, and most importantly, establish contingency plans. Factoring in circumstances that are beyond your control is essential. Such circumstances include those that emerge from external surroundings, with the potential to impact the cost of supplies and labor and to cause product, service, and resource shortages.

For example, the currency exchange rate may fluctuate. The rate of exchange today may not be the rate of exchange tomorrow. This means that the exchange rate at the initiation of the project might be different from the exchange rate during the later phases. Additionally, too many project controllers are caught unprepared by vendors who fail to meet their obligations. Therefore, it is important for a project controller to ensure that the vendors involved in the project deliver on their promises, and to have a backup plan in case they fail to deliver. Involving all project stakeholders in the budgeting process, as well as vetting vendors and suppliers, can go a long way in helping the project controller set a more realistic budget that can be met even in unforeseen circumstances.

3. Develop relevant Key Performance Indicators (KPIs)

As a successful project controller, you cannot manage a budget without the establishment of KPIs. It is these KPIs that play a very important role in helping you and the rest of the team ascertain how much has been spent on the project activities, and how the initial budget differs from the current expense status.

Some of the most commonly used KPIs that are essential for proper project budget management include:

- ✓ **Actual cost (AC)**: this is also referred to as Actual Cost of Work Performed (ACWP). This demonstrates the amount of money that has been spent on the project to date.

- ✓ **Cost variance (CV)**: this indicates whether the estimated cost of the project is above or below the established baseline.

- ✓ **Earned Value (EV)**: this is also referred to as Budgeted Cost of Work Performed (BCWP). This demonstrates the approved budget for already performed project activities as of the date of reporting.

- ✓ **Planned Value (PV)**: also referred to as Budgeted Cost of Work Scheduled (BCWS). This is the cost estimated for the project activities scheduled as of the date of reporting.

- ✓ **Return On Investment (ROI)**: this demonstrates the profitability of the project and whether the benefits have surpassed the costs.

4. The "Three Rs"

This stands for **Revisit, Review,** and **Re-forecast**. Always bear in mind that if you let the project run without a budget or any re-forecasting, you are preparing to fail. To ensure that you prevent the project from getting too far out of hand, it is important to conduct frequent budget oversights. A 10% budget overrun is much easier to correct than a 50% overrun. If you fail to keep an eye on your budget and re-forecast appropriately, the 10% overrun can easily become a 50% overrun! However, with frequent project reviews, you will be able to keep the project on track.

In the same way a budget needs to be frequently revisited to ensure the project is on the right track, so does the usage of project resources. This is because most of the people working on the project play a huge role in contributing to the cost of the project. As a project controller, it is your role to review the number of people that are working on the project, as well as to assess weekly the future resource needs of the project. In doing so, you are ensuring that all resources are being fully used, and that the right resources are in place for the rest of the project's activities.

Additionally, you have to constantly re-visit the resource forecast to keep the budget on the right track. Some of the leading causes of budget overruns are scope creeps. This means that so much unplanned work has found its way into the project that the number of billable hours increases significantly, causing a loss of budget control. As a project controller, ensure that you are exercising care and caution with scope management. You can do this by simply requiring change orders for work that is not covered by the initial requirements of the project. These change orders are vital in authorizing additional funding to cover the cost of any extra work, and keeping the project aligned with the now-increased budget.

5. Keep everyone informed and accountable

One of the most important parts of sticking to the budget is making sure that all the members of the team are constantly aware of the budget status. Ensure that you keep the team members up to date with budget forecasts. Why is this important? When your project team is well informed, they are empowered to take ownership of the project. By ensuring that the team knows the status of the budget, you are encouraging them to monitor their project charges, which makes it less likely that they will charge gray area work hours to the project. "Gray area hours" refers to hours during which team members worked, but lacked certainty of what they were working on.

The budget is a living part of the project. It is something that any successful project controller regularly reviews with project teams and stakeholders. Simply put, a project controller who watches the budget closely and involves the stakeholders will enjoy a happy project management, and will experience greater success both in the project and in their career.

Defining Proper Project Control

In previous sections, we discussed the traits of a successful project controller. But what are the traits of successful project control? In this section, we will discuss proper project control practices.

Yes, all project leaders in any organization strive to ensure that their projects are successful, profitable, and have the capacity to contribute to the growth of the company while attracting the best clients and future projects. Indeed, many professional firms believe that they have good project control in place, and that they are able to deliver profitable results. However, if you look closely at their projects, you will realize that many have not turned out to be truly profitable. In fact, many firms today are operating blindly. Many organizations are managing their enterprises without the necessary information required to keep up with the competition in the modern day business arena. Furthermore, many of the firms that claim to have real insights are only acting on a couple of weeks of data!

If you cannot predict what lies in the future, you cannot act until it is too late to make the required changes. Do you have the insights required to predict the expected outcome of the project during any phase of its life cycle? If you do, you are one of the few who already know how foresee potential risks and make adjustments before the project becomes a disaster. If you do not, this section will offer you some tools for improving your project control and profitability.

A challenge for many project controllers is the fact that they have to assess several different business arenas in order to get a good overview of the projects they control. Fortunately, there is a solution for this. A single workspace can be created where project professionals can easily monitor various phases of their projects. It is important to have ERP solutions to boost the profitability of the project and minimize possible write-offs. These give you real-time control over the project, as well as insights that will enable you to make relevant, fact-based, and timely decisions.

Principles of Project Control

The cornerstone of any organization is profitable projects. It is the profitability of the project that will determine profit margins. This is something that is very important, especially if you run fixed-price projects, in which budget overruns affect the bottom line directly. Project write-offs often lead to billable hours that are in turn written off, resulting in company losses because they are not billed to the client. So, you ask, how do I improve project control and profitability? The following are four basic project control principles:

1. Clarify project baseline

Setting up the project according to a defined structure plays an important role in effectively seeing the project through from initiation to closure. This helps ensure that the scope of work is defined and mirrored in the baseline budget, serving as a cost control point. In other words, it is important that each project has a baseline budget that serves as an estimate from the time the contract starts. As an advanced project professional, you can calculate a separate working budget in addition to the baseline budget. This helps you handle the true costs rather than just the contracted expenses. In addition to improving the functionality of a budget, the ERP provides a revision history to track any changes made to the budget.

2. Control project cost

Throughout the life cycle of the project, it is important to remember that project control is all about controlling the actual cost of the project, using the cost estimated in the baseline budget as a reference. Therefore, it is important for you to ensure that the right resources are made available and locked into the project. It is also important that you control all other project costs such as time, purchases, and other expenses, so that you have the overall cost under control. The best way to control costs is to monitor the hours spent. It is your responsibility as a project controller to ensure that the number of billable hours worked by personnel fall within the project's budget. To optimize cash flow and maximize revenue, make sure that all customers are invoiced on time.

3. Control project scope

It is your role as a project controller to effectively manage all scope change requests, in order to prevent a reduction in the profitability of the project. For instance, if you and the team engage in activities that were not initially reflected in the baseline budget, you may end up incurring extra costs that were not planned for at the start of the project. This can have a negative impact on the profitability of the project. Therefore, it is important to integrate proactive change management procedures, and update the Estimation To Completion (ETC). This will not only offer a baseline for change management, but also allow the project to adapt to scope evolution.

4. Proactively evaluate progress

In order to prevent overruns and delays, it is important to continually evaluate the progress of the project. Using the ETC to manage actual and upcoming costs will help you get a broad perspective of what the costs will be at the completion of the project. You will then compare this cost to the baseline budget. This helps you to estimate the time remaining before the project comes to an end, and enables you to make adjustments before it is too late. The best way to do this is with a resource planning solution for constantly and continuously updating the resource plans. This helps you see the latest cost estimate on the ETC. An ERP that is tailor-made for your organization is a great tool for managing and controlling the profitability of each project.

Five Effective Project Control Practices

In this section, we will discuss practices that are based on the principles detailed above. Effective project control ensures that the work activities of the project are managed well during each part of the implementation phase. As discussed earlier, project control is a management function, and its purpose is to achieve defined project goals and expectations within the scheduled time. In the past, project control involved three major high-level processes that the management team was required to carry out:

- ✓ Setting high standards for the project
- ✓ Measuring work performance and progress
- ✓ Taking corrective actions

Today, these processes have been further divided into five specific practices: holding meetings, performing quality control, tracking progress, responding to changes, and managing risks and issues.

1. Holding meetings

Assembling the project leadership ensures that the project continues to run according to defined objectives. At a meeting, it is the responsibility of the project controller to provide an overview of the current tasks of the project, describe current goals and issues, and identify effective communication channels.

Every meeting should begin with an agenda. The manager writes the agenda, and then shares this information with the rest of the team. Conducting meetings is a very important step in accomplishing tasks of the project and in controlling project processes. It is during meetings that the project controller can assign or reassign roles and responsibilities to participants. The project controller also offers executive direction of the project to the participants and other stakeholders, notifies the stakeholders of the current state of the project, and guides the team. Whenever there are internal or external issues within the project, it is the responsibility of the manager to make executive decisions pertaining to any corrective action. Finally, during these meetings, the project controller establishes and conducts reviews of project success, using set criteria.

2. Performing quality control

Quality control makes it possible to determine whether the product is complete and developed in line with expectations. Quality control is all about identifying defined standards, business expectations, and established product requirements, among other factors. This process starts when the project is initiated, and continues throughout the life cycle. It continues until the final product has been handed over to the customer. Activities and tasks at any stage of the project life cycle must be signed off properly, for the purpose of project continuity. Some of the quality control tasks that the project controller has to perform include:

- ✓ Creating a quality control schedule that determines when quality control takes place in each phase
- ✓ Developing an agenda to determine the key tasks of the people taking part in the quality control process
- ✓ Assigning reviewers whose main focus is to assess commitments, responsibilities, and objectives, among other factors
- ✓ Designating other roles, such as facilitation or authorship roles
- ✓ Documenting and recording all actions and decisions that the project team makes throughout the quality control process
- ✓ Performing follow-ups and notifying key stakeholders of the status of the project once the quality control process is over

3. Tracking progress

This practice involves monitoring, measuring, and controlling the project progress. The main aim of this practice is to ensure that the work of the project is conducted according to a defined schedule. The role of the project controller is assess progress throughout every phase, making sure that activities run smoothly in the right direction. How do project controllers track progress? These are the steps they follow:

- ✓ First, the project controller captures task performance data. This data includes actual start and completion dates of tasks, actual hours logged for the various project tasks, and latest estimated duration of tasks in hours, among other data.
- ✓ Second, the project controller updates the schedule with the actual task performance data. This allows the controller to estimate the remaining cost of the project, and to update the cost estimates along with the actual costs incurred during a specified period.
- ✓ Third, he or she will capture the costs incurred in non-staff activities, and then consider re-planning the project work for a given stage. This re-planning often

involves updates that will help in estimating the schedule and cost. This is critical in ensuring that there are sufficient staff members available and that re-assignments of any additional resources are made.
- ✓ Fourth, the project controller will measure the performance of team members, and determine any possible issues that could be contributing to low performances. This way, personnel can be motivated properly, and corrective actions can be taken to eliminate any issues relating to performance.

4. Responding to changes

Controlling change means that any additional work introduced is well-defined and implemented. By effectively responding to changes, the project controller ensures that such aspects as scope, cost, and time remain in line with the expected status. To this effect, he or she must receive and review any change requests. This enables a better understanding of changes proposed as matters of priority. As a project controller, you can also delegate these changes to competent members of the team who can come up with practical alternative solutions. Once alternatives are assessed, you can respond to the change request accordingly. When a change request is approved, an action plan is created to implement these changes. The implementation timeline should be defined, in order to ensure that quality progress is achieved within the specified window.

5. Managing risks and issues

The main aim of this practice is to resolve any issues that arise and jeopardize success during the course of the project. This practice often involves a series of steps. These steps include identifying the issue, assessing its impact, establishing corrective action, taking action, and then tracking progress. As a project controller, you have to manage these issues well, so that the project runs as planned. Therefore, you have to identify and keep record of any issues that could potentially affect the project. Keep an issue log that specifies and describes each issue. The information you should log includes the type of issue, the status, personnel involved, and priority. Additionally, you must adequately assess the impact of the issues relating to cost, time, and scope, in order to better come up with resolutions.

Indeed, project control is a complex phenomenon. However, when you implement all these principles and practices, you can effectively and successfully lead a project to its set goal.

CHAPTER SIX

Project Management Methodologies

It is your role and duty as a project professional to help your organization effectively implement its projects, while reducing risks. This role requires much more than simply recognizing the priorities of the organization. It is important that you also have an in-depth understanding of Project Management Methodologies (PMMs), in order to improve the likelihood of your organization's success with projects.

Here are some of the most popular PMMs that you can put into practice today. Make sure to evaluate which method applies best to a particular project. These include:

- ✓ **Waterfall** PMM

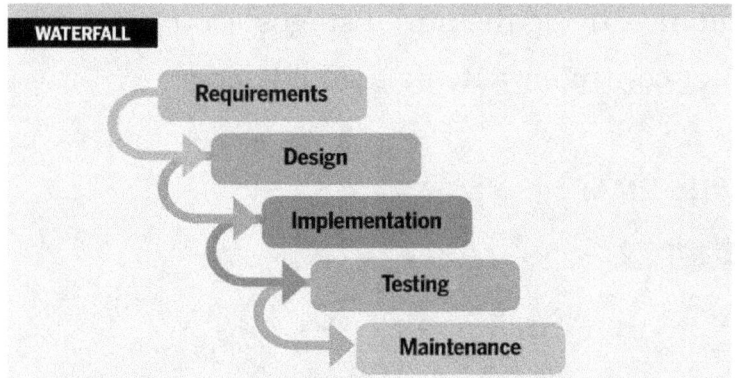

Figure 10. Thinkstock - Waterfall PMM.

This is one of the mainstays of project management methodologies; it has been around for several years. It involves a sequence of **static steps**, and is applicable to many industries, especially the software development industry. The steps are requirement analysis, design, testing, implementation, and maintenance. These steps require that tasks be executed in a specific order. The waterfall methodology permits for increased control throughout the phases of project management. However, it is important to note that these phases can be quite inflexible, which can cause problems if the scope of the project changes after it is already underway. This means that this methodology requires a more formal planning stage to make sure all requirements are captured beforehand. This, in turn, reduces the chances of missing key information and requirements during the initial stages. In the waterfall PMM, the project controller acts as a gate keeper.

✓ **Agile** PMM

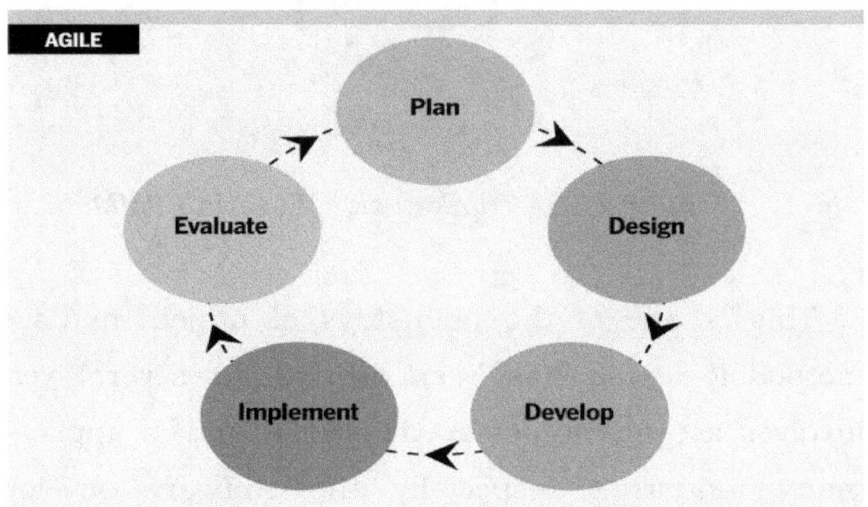

Figure 11. Thinkstock Agile methodology.

This methodology employs a significantly different approach from the waterfall PMM. The agile PMM was specifically developed for projects that require high levels of flexibility and speed. Accordingly, this methodology is comprised of short delivery cycles, referred to as "sprints." It is best suited for projects that require less control, and demands real-time communications within a self-motivated team. Agile methodology is highly interactive, permitting for rapid adjustments throughout the project life cycle. This methodology is commonly used in software, web development and related projects, due to the ease with which issues and modifications can be identified and made during early project phases. One of the best things about this methodology is the fact that it permits repeatable processes, offers reduced risks, and boosts immediate feedback. This all contributes to fast turnaround times and reduced levels of complexity. Agile PMMs enable the project controller to move forward with one part of the project, while holding back another to make changes.

One type of agile PMM is known as **scrum**, like the rugby term. Scrum sessions are interactive in nature, and play an important role in prioritizing tasks. It is the responsibility of a "scrum master" to facilitate sessions, rather than of a project manager or controller. In scrum, and in other agile methodologies, the project controller is only in charge of monitoring budgets and offering administrative support. In the scrum approach, small teams are assembled to focus independently on specific tasks. They then meet with their scrum master to evaluate the project progress, results, and to re-prioritize other backlogged tasks.

✓ **Hybrid** PMM

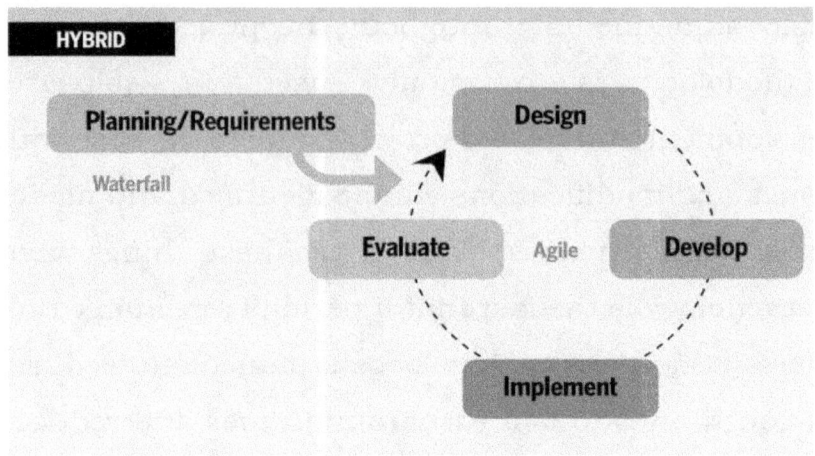

Figure 12. Thinkstock- Hybrid PMM.

Many project teams tend to favor either the waterfall or agile PMM, so the hybrid approach combines the advantages of both. This combination of benefits makes this methodology even more suitable than the others. The planning and requirement phases follow the waterfall approach, while the design, development, implementation, and evaluation phases follow the agile approach.

In the hybrid method, the project controller plays the gatekeeper role during the waterfall-inspired planning phases, and acts as budget monitor and administrative support during the agile-inspired phases.

When assessing PMMs from the end-user's point of view, the project controller must be able to curb any roadblocks that result in product limitations. Customization and data integration are two PMM processes that can present difficulties.

✓ Customization

Despite the fact that customization is a component of many tools, it is not easy to apply. Even when the customization process is straightforward, it can often be limited by the general logic frame of the PMM.

✓ Data integration

This is often required when multiple platforms are used for different needs of the organization. For example, a project might be managed through Project Management System, while the methodology is managed through a different tool, and at a much higher level. However, although they complete each other, these tools can be in competition from the perspective of data integration.

How, then, do these PMMs influence the role of a project controller?

In today's economy, most work is specialized, is dependent on technology, has short deadlines, is mobile, and requires increased collaboration of resources and social skills. In this regard, project management should focus on improving product value, making hands-on decisions, and adapting to changes.

It is the responsibility of project professionals to recognize that PMMs enable adaptation to change. PMMs are the most effective and favored ways by which project professionals manage change, and failure to use PMMs may jeopardize the overall effectiveness of a project. The project controller is indispensable when it comes to change management. He or she is at the center of driving necessary transformations. It is through the use of PMMs that the controller can efficiently prepare for change and put contingency plans in place.

The keys to selecting the most relevant PMM are that project managers ensure that their controllers understand project requirements, know what resources are available, and can maintain suitable control of resources throughout the project. The project controller must be adaptable, meaning that they are able to discern the merits of a potential PMM while already working in the context of the project.

Additionally, when an organization is considering the replacement of an institutionalized PMM, the project controller must be able to understand and advise the project manager on the importance of context and how it is mirrored in the incumbent methodology. With this in mind, the project controller can help the project manager make an informed decision. The project controller's role is to always be on the lookout for environmental factors that may negatively impact the project, and advise the project manager. In turn, the project manager will be able to understand and manage potential compromises to the PMM's function.

CHAPTER SEVEN

Common Project Management Mistakes

We will use this chapter to present common mistakes in project management that are critical to avoid. Countless projects have been mismanaged in the past, and many continue to be mismanaged today. You may wonder how that can be the case, as there is so much project management software available. The truth is that, even very good project professionals with high-end software can end up making mistakes as they wrangle big and complex projects.

Most projects often end up taking longer than scheduled, and therefore, costing more than planned. While no two projects are exactly alike, the issues that put projects in jeopardy are often quite similar. For instance, mistakes often occur when a project is bombarded with too many change requests. Some common mistakes that you should be aware of include:

- ✓ Failure to meet with the whole team to set goals up front

According to many project managers, it is crucial for the entire project team to know their respective roles, responsibilities, and required deliverables right from the very beginning. Therefore, it is advisable for the project controller make sure that the project manager convenes a kickoff meeting with all project stakeholders. It is this kickoff meeting that is important in helping define expectations of the project team. This ultimately helps the project team feel independent and organized from the get-go. Additionally, such meetings often instill in the stakeholders a sense of accountability and ownership of the project.

- ✓ Failure to break down large projects into smaller, specific pieces

Most of the time, the projects that organizations engage in are large in scope. It can be highly complicated to conduct and monitor the activities of the project as a whole. It is important to avoid overwhelming the project team, and taking the time to understand every single facet of the project goes a long way. Breaking up the project into smaller, more manageable chunks often helps the project team feel much more comfortable when tackling project activities. In other words, this helps them feel a sense of confidence that they can achieve success in what might otherwise appear to be an impossible task, if the project was viewed in its entirety. As the project controller, you can work with the project manager to break the project into smaller tasks for each team member to deliver, and to assign the tasks according to members' specific areas of expertise.

Failure to prioritize tasks and projects

Many organizations have projects that run concurrently both within and outside of the institution. Often times, project controllers concentrate too much on projects of low priority, while letting those with high priority slip. This is why project managers have to give team members adequate awareness of the priority of each assigned tasks. This way, members will prioritize tasks of greater importance and impact. They must also know when priorities change, and how to re-prioritize their tasks. The project controller should guide and help the project manager communicate task priorities with a high degree of clarity and accuracy, in order to save time.

Forgetting that project management is also people management

It is an alarming truth that many project professionals tend to get so bogged down with the technical details of project activities that they forget about the people who are actually involved in these activities. It is important to understand that it is not just about the delivery of work, but also about the team delivering the work. Failing to manage team members well—for instance, by micromanaging them—often can result in delays, cost overruns, and negative impacts on the quality of work. To avoid this, the project controller must help everyone involved understand perfectly how and why their role is important to the success of the project. This can be achieved by performing periodic check-ins with team members, sponsors, suppliers, and executives of the project, among other stakeholders. This ensures that all the people involved, whether directly or indirectly, share the same vision for how the project is to be managed.

Failure to communicate regularly with the team members

While no one refutes the importance of constant communication in project management, once the project is underway, it is all too easy for professionals to forget to meet with the project team and to provide updates to key stakeholders. In order to prevent this from happening, the project controller should help the project manager establish a foundation of regular meetings. They should plan important factors, like who should attend. It is important to involve the right number of people. If there are too many people in the meeting, progress may be bogged down if everyone feels the need to comment at every single turn. Too few people involved means that not enough project workers receive adequate information.

Allowing changes get out of hand

One of the most common mistakes that project professionals make is poor handling of scope creeps. This is something that is difficult to manage because, just as the name suggests, the changes "creep" up on you. When there are additional requests for features that were not initially budgeted for, a severe strain is put on the budget, and the focus of the project vision is affected. Without proper control established by the project controller, this can significantly impact the success of the project.

In order to curtail scope creeps, strong project management practices must be in place, as well as product ownership. When considering the addition of new features to the project, you have to be able to answer a number of questions: do these new features align with the vision of the project? Do the changes add to the value of the end product? Are these changes critical, or at least helpful to add? These questions put the changes in perspective, enabling you to redefine project goals and to identify potential success factors. You must make certain that change requests for additional features that are not in line with overall project objectives do not threaten set timelines and budgets.

Failing to use a project management tool

Remember that most project management tools offer great visual representations of the project status. In order to keep projects on track, it is paramount that these tools are used to ascertain the project's progress. This way, you can easily tell if the project is performing sluggishly, or if it's current pace will likely lead to completion by the required deadline. This helps identify potential opportunities to increase efficiency, like by quickly spotting potential problems. Additionally, it is important to document project activities and milestones on a weekly basis. When something of great importance crops up, the project manager and controller should update the documentation within 24 hours. This way, everyone on the project will have access to accurate, current information.

Not adjusting the course of the project when things go wrong

As we all know, sometimes, despite our best efforts, things do not go according to plan. In such situations, there is a fear that any efforts exercised to salvage the situation will be futile. This is why it is so important for a project controller to create a vehicle for transparent and accurate reporting, which provides stakeholders with the right information for timely decisions. If a project is too strategically important to abandon in the face of difficulties, proper reporting practices can enable simple adjustments to budget, resources, and expectations, which can help a project change course and become much more successful.

CHAPTER EIGHT

Take Control of Your Career

Traits of an Exceptional Project Controller

Did you know that to be a great project controller, you have to be a strategic business partner? Being a strategic business partner ensures that you are fully invested in the success of the organization. There are certain traits that you, a successful project controller, has to have in order to stand out from the crowd. The truth is, every project controller knows how to control projects within the budget and timeline, provide reports, and ensure all stakeholders' requirements are met consistently. However, not all project controllers understand that to be great, they must go above and beyond. You must not only keep projects within the defined scope, time, cost and quality, but also be accountable and strategic in ensuring the success of the organization.

Therefore, if you are looking for ways in which you can stand out and advance your project management career, it is these traits that will effectively take you to the next level. These traits, if coupled with the necessary technical skills, will place you in higher demand as a project professional. You will be better adapted to the ever-changing demands and dynamics of projects, and in a better position to prioritize the needs of the stakeholders above all else. Traits of a stand-out project controller include:

✓ Being a strategic partner

Being a strategic partner means you not only offer high-end leadership skills, but also offer significant technical management skills that provide the organization with the advantage they need to seize success. Today, there are quite a number of factors that may influence the success of a project, both internally and externally. These factors include the **triple bottom line**, as well as legal, legislative, international, cultural, and remote project issues. Triple bottom line factors entail economic, ecological, and social outcomes. It is factors such as these that often create additional complexities with which to contend. Therefore, if you do not have an in-depth understanding of how a project fits into the overall strategy of the company, you may end up limiting your chances of effective delivery. In particular, the Executive Project Management Office (EPMO) and portfolio PCOs have to pay close attention to these strategic connections, in order to promote the success of projects, programs, and portfolios.

- ✓ Encouraging and recognizing valuable project contributions

It is important for everyone to understand that the success or failure of a project does not fall on the shoulders of just one individual. Effective project leadership is strongly impacted by contributions of the project team. Make sure to share credit for work well done with all the other members of the team. This kind of recognition goes a long way in encouraging all members to contribute and participate at the highest level possible. Furthermore, delegate responsibility! Instead of trying to be a jack-of-all-trades, you can take advantage of the knowledge and skills of the other members of the team. These simple tactics are highly effective in increasing the chances of achieving project goals.

- ✓ Respecting and motivating stakeholders

The success of a project often relies on the effectiveness of the communication with, and the influence of, all the stakeholders. You need to find ways to motivate workers over whom you may not have direct influence, but who can make or break the project. The best way is to simply instill in them a strong sense of confidence, especially when there is a need for scope change in the project. In order to earn the respect of the project team members, sponsors and all other stakeholders, it is important that you accord them the respect they deserve at all times. It is almost always impossible for the project to progress in the right direction without respect and motivation, especially from workers and financers of the project.

- ✓ Having accountability and integrity

As we have discussed, something to bear in mind is that not everything will go as well as planned. In other words, mistakes should be expected. To this effect, it is critical for the success of the project to accept when we are wrong and to learn from the experience. It is all about being accountable for our decisions and actions. Accountability sends a strong, positive message to the stakeholders of the project. The hallmarks of a successful and effective project controller are integrity and good judgement. Expressing these characteristics inspires a high degree of confidence, and demonstrates competence. If you lack these traits, you may not be able to get people to follow your lead. Poor integrity and judgement are often reasons for failure in project leadership.

✓ Working in the gray

Many of the attributes that we have discussed speak for themselves, but what truly sets a leader apart is their ability to "work in the gray." This is something that most people overlook, yet it is the most important attribute irrespective of the size, organization, complexity, and type of project. Every project will have gray areas that require navigation. Some of these gray areas include external constraints, conflicts, ambiguities, and remote project limitations. These complexities will almost always be encountered. Therefore, you have to be able to approach these complexities and changes in a wise manner to ensure that you can discern when a project is in trouble. The best project controllers face delays and budget cuts without being rattled, have high-end technical and people skills, and can work effectively in the gray.

✓ Being fully invested in success

In order to be effective and successful, you have to strongly believe in yourself and in what you do. You must be willing to see the project through, from the initiation stage all the way to closing. It is this mindset that will help you achieve good results throughout the project. You must be fully engaged in all aspects that relate to the project, especially the activities and the people involved. However, in order to maintain the integrity and satisfaction of all stakeholders, you must avoid overextending yourself. In other words, you have to balance the needs of clients and workers in order to satisfy everyone involved.

✓ Taking career control

Many project managers and project controllers do not do enough to advance their careers and know-how in the project management field, in spite of current changes and the need for more project controllers' skill sets across a wide range of organizations and projects. This is especially the case this year, in which almost one third of project professionals are predicted to have a 0% salary increase. The biggest priority for project controllers and managers should be to improve their jobs by earning prospects and technical know-how. In other words, for every career opportunity out there, it is important for project controllers to aspire to be among the top 10% of candidates. There are a number of strategies to consider when trying to attain career growth:

- Have a plan

Many people often think they can simply settle for doing a good job. The truth is, you can do more than you realize, and you should always be prepared for unexpected setbacks: organizations around the world often make the decision to cancel projects, especially when projects are too similar to those already running, in order to introduce new and different ones. If you do not have a proper plan, you just might be among those left out in the cold.

- Be informed

Often, external influences can affect your career. If you do not stay up-to-date with the current trends, changes, and dynamics of the business world, there is a high risk that you will be caught off-guard. For instance, based on the latest benchmark report, almost a quarter of project managers were not aware of the chartered status in the profession. Of more concern is that more than a third were not aware of the year's changes to off-payroll legislations, which have a huge impact on project managers operating as limited companies. Therefore, the things in your profession about which you lack information may end up affecting your career. It is important to acquire the latest training, and to do research to build your knowledge and relevance to the field.

- Understand gaps

In most cases, project professionals delegate certain duties in order to concentrate on other work. As a project controller, take advantage of this by taking up duties that will help bring out your strengths and give you more experience. Additionally, it is important that you choose to widen your skill set by taking on something new. As a good project controller, you need to know enough about any given phenomenon to at least be able to talk about it in an intelligent manner.

- Take control of your development

You are a project controller today, but tomorrow, you might desire to be something more. You may want to be a project manager, program manager, or even a portfolio manager. However, these prospects might not materialize under your current employer. In such a case, it is important for you to consider self-funding to help further achieve your prospects. In other words, if progressing might mean that you move to another organization, you should start reasoning like a chess player, thinking a few moves ahead.

- Understand the bigger business picture

Today, there are numerous organizations engaging in portfolio management. As a project controller, you need to understand where in the portfolio your project fits. The truth of the matter is that your project will be affected if another takes precedence. You have to understand the business in order to take up a role outside project control, in programs or portfolios. However, it is important to note that there is a big gap between these roles. You will have to develop your business acumen, in addition to both technical and soft skills.

- Network, network, network

A functional network is one that is made up of two-way relationships. It is important for you to take the time to recognize the importance of your network and the source of your career breaks. In most cases, they come from people you already know and trust. It is through personal networks that you can advance your career in ways that would not be possible on the open job market.

- Develop your people skills, leadership, mentorship, and coaching

As a good project controller, it is important to have a mentor, as well as to coach other project professionals. If you take the time to mentor and be mentored, you can develop your skill set as well as those of others. Additionally, you can give back by doing presentations, joining a group, or volunteering you skills to a charity. It is through such acts of service that people will begin to recognize you as a valuable representative of the project management profession.

- Recognize the importance of work-life balance

One of the mistakes that a lot of project professionals make is burying their heads in work, work, and more work! Let's face it, project management and control can be very stressful. However, the best way to deal with stress is to plan well, and sometimes even take several months away. A career break is very important, even required, for one to reflect and come back to work refreshed, possessing a different perspective. Despite what might think, a career break often boosts your salary-earning potential significantly!

One Last Word

In summary, project control is concerned with saving time and money during project planning and execution.

This is something that you have to constantly strive for in order to survive and grow in the industry. If you were afraid of how project control is vitally concerned with the use of project management software, I bet the fear is gone now! According to the omniscient PMBOK, project control is all about data-gathering, management, and analytical processes that help predict, understand, and constructively use time and money in a project. It is through effective and efficient communication that information flows, helping project controllers make the right decisions within their dockets.

Now that you know what project control is all about and how you can spread your wings and soar in this profession, consider what would happen without proper control practices: just the opposite of control, which is chaos! In other words, if you ignore the project control tips and tricks presented in this book, you are simply getting ready to fail.

It is our hope that you take full advantage of all the information presented to you in this book. It is this information that will help you thoroughly incorporate proper and effective project control practices into your project management career. Despite being expensive, time consuming, and labor intensive, performing detailed and well-adhered to practices will be well worth the effort, and customers will take notice. What more do you need?

References

Anyosa Soca, V. & Rojas, J. (2007). PMRS: project management reinforcement and support process. Paper presented at PMI® Global Congress 2007—North America, Atlanta, GA. Newtown Square, PA: Project Management Institute.

Barkley, B. T. (2004). *Project risk management (Project management)* (Vol. 1). McGraw-Hill Professional.

Bowenkamp, R. D., & Kleiner, B. H. (1987). How to be a successful project manager. *Industrial Management & Data Systems, 87*(3/4), 3-6.

Duncan, W. R. (1993). The process of project management. *Project Management Journal, 24*(3), 5-10.

Frame, J. D. (1999). *Project management competence: Building key skills for individuals, teams, and organizations* (p. 232). San Francisco, CA: jossey-Bass.

Hayes Munson, K. A. (2012). How do you know the status of your project?: Project monitoring and controlling. Paper presented at PMI® Global Congress 2012—North America, Vancouver, British Columbia, Canada. Newtown Square, PA: Project Management Institute.

Homer, J. L. (2004). The role of project control systems in facilitating and measuring project success. Paper presented at PMI® Global

Congress 2004–North America, Anaheim, CA. Newtown Square, PA: Project Management Institute.

Kerzner, H., & Kerzner, H. R. (2017). *Project management: a systems approach to planning, scheduling, and controlling.* John Wiley & Sons.

Larson, E. W., & Gray, C. F. (2015). A Guide to the Project Management Body of Knowledge: PMBOK (®) Guide. Project Management Institute.

Lester, A. (2006). *Project management, planning and control: managing engineering, construction and manufacturing projects to PMI, APM and BSI standards.* Elsevier.

Lock, D., & Wagner, R. (Eds.). (2016). *Gower Handbook of Programme Management.* Routledge.

Morris, P. W. G. (1980). The use and management of project control systems in the 80s: a proposal. *Project Management Quarterly, 11*(4), 25-28.

Project Processes Project Management Institute (2013). A Guide to the Project Management Body of Knowledge (PMBOK®Guide). Project Management Institute, Inc. (5th ed., pp. 50).

Rathore, A. (2010). The growing importance of EPMO (Enterprise Project Management Office) in today's organizations. W *ipro*

Technologies [http://www. projectsmart. co. uk/docs/the-growing-importance-ofepmo-in-todays-organisations. pdf](12.01. 2013).

Reich, B. H., & Wee, S. Y. (2006). Searching for Knowledge in the PMBOK® Guide. *Project Management Journal, 37*(2), 11-26.

Reiss, G. (2006). *Gower Handbook of programme management.* Gower Publishing, Ltd.

Rosenau, M. D., & Githens, G. D. (2011). *Successful project management: a step-by-step approach with practical examples.* John Wiley & Sons.

Rosenking, J. P., Forgang, C., Jacobs, B., & Bergin, M. E. (2002). Project Management planning, process and procedures. Paper presented at Project Management Institute Annual Seminars & Symposium, San Antonio, TX. Newtown Square, PA: Project Management Institute.

Ross, D. W. & Shaltry, P. E. (2005). The new PMI Program management standard and Portfolio management standard—impact on the profession—a preview. Paper presented at PMI® Global Congress 2005—North America, Toronto, Ontario, Canada. Newtown Square, PA: Project Management Institute.

Rozenes, S., Vitner, G., & Spraggett, S. (2006). Project control: literature review. *Project Management Journal, 37*(4), 5-14.

Thamhain, H. J. & Wilemon, D. L. (1986). Criteria for controlling projects according to plan. *Project Management Journal, 17*(2), 75-81.

Yosua, D., White, K. R. J., & Lavigne, L. (2006). Project controls: how to keep a healthy pulse on your projects. Paper presented at PMI® Global Congress 2006—North America, Seattle, WA. Newtown Square, PA: Project Management Institute.

www.ingramcontent.com/pod-product-compliance
Lightning Source LLC
Chambersburg PA
CBHW070645220526
45466CB00001B/302